OPERATION OF RESTRUCTURED POWER SYSTEMS

THE KLUWER INTERNATIONAL SERIES IN ENGINEERING AND COMPUTER SCIENCE

Power Electronics and Power Systems

Series Editor

M. A. Pai

Other books in the series:

TRANSIENT STABILITY OF POWER SYSTEMS: ***A Unified Approach to Assessment and Control***
Mania Pavella, Damien Ernst and Daniel Ruiz-Vega, ISBN 0-7923-7963-2
MAINTENANCE SCHEDULING IN RESTRUCTURED POWER SYSTEMS
M. Shahidehpour and M. Marwali, ISBN: 0-7923-7872-5
POWER SYSTEM OSCILLATIONS
Graham Rogers, ISBN: 0-7923-7712-5
STATE ESTIMATION IN ELECTRIC POWER SYSTEMS: ***A Generalized Approach,*** A. Monticelli, ISBN: 0-7923-8519-5
COMPUTATIONAL AUCTION MECHANISMS FOR RESTRUCTURED POWER INDUSTRY OPERATIONS
Gerald B. Sheblé, ISBN: 0-7923-8475-X
ANALYSIS OF SUBSYNCHRONOUS RESONANCE IN POWER SYSTEMS
K.R. Padiyar, ISBN: 0-7923-8319-2
POWER SYSTEMS RESTRUCTURING: ***Engineering and Economics***
Marija Ilic, Francisco Galiana, and Lester Fink, ISBN: 0-7923-8163-7
CRYOGENIC OPERATION OF SILICON POWER DEVICES
Ranbir Singh and B. Jayant Baliga, ISBN: 0-7923-8157-2
VOLTAGE STABILITY OF ELECTRIC POWER SYSTEMS, Thierry Van Cutsem and Costas Vournas, ISBN: 0-7923-8139-4
AUTOMATIC LEARNING TECHNIQUES IN POWER SYSTEMS, Louis A. Wehenkel, ISBN: 0-7923-8068-1
ENERGY FUNCTION ANALYSIS FOR POWER SYSTEM STABILITY, M. A. Pai, ISBN: 0-7923-9035-0
ELECTROMAGNETIC MODELLING OF POWER ELECTRONIC CONVERTERS, J. A. Ferreira, ISBN: 0-7923-9034-2
MODERN POWER SYSTEMS CONTROL AND OPERATION, A. S. Debs, ISBN: 0-89838-265-3
RELIABILITY ASSESSMENT OF LARGE ELECTRIC POWER SYSTEMS, R. Billington, R. N. Allan, ISBN: 0-89838-266-1
SPOT PRICING OF ELECTRICITY, F. C. Schweppe, M. C. Caramanis, R. D. Tabors, R. E. Bohn, ISBN: 0-89838-260-2
INDUSTRIAL ENERGY MANAGEMENT: ***Principles and Applications***, Giovanni Petrecca, ISBN: 0-7923-9305-8
THE FIELD ORIENTATION PRINCIPLE IN CONTROL OF INDUCTION MOTORS, Andrzej M. Trzynadlowski, ISBN: 0-7923-9420-8
FINITE ELEMENT ANALYSIS OF ELECTRICAL MACHINES, S. J. Salon, ISBN: 0-7923-9594-8

OPERATION OF RESTRUCTURED POWER SYSTEMS

by

Kankar Bhattacharya

Math H.J. Bollen

Jaap E. Daalder

Chalmers University of Technology

KLUWER ACADEMIC PUBLISHERS
Boston / Dordrecht / London

Distributors for North, Central and South America:
Kluwer Academic Publishers
101 Philip Drive
Assinippi Park
Norwell, Massachusetts 02061 USA
Telephone (781) 871-6600
Fax (781) 681-9045
E-Mail <kluwer@wkap.com>

Distributors for all other countries:
Kluwer Academic Publishers Group
Distribution Centre
Post Office Box 322
3300 AH Dordrecht, THE NETHERLANDS
Telephone 31 78 6392 392
Fax 31 78 6546 474
E-Mail <services@wkap.nl>

Electronic Services <http://www.wkap.nl>

Library of Congress Cataloging-in-Publication Data

A C.I.P. Catalogue record for this book is available
from the Library of Congress.

Printed on acid-free paper.

Printed in the United States of America

Contents

List of Acronyms xi
Preface xiii

CHAPTER-1: DEREGULATION OF THE ELECTRICITY SUPPLY INDUSTRY

1. **Introduction** 1
2. **What is Deregulation?** 2
   2.1 Different Entities in Deregulated Electricity Markets 4
3. **Background to Deregulation and the Current Situation Around the World** 6
   3.1 Industrialized Countries 6
   3.2 Developing Countries 15
4. **Benefits from a Competitive Electricity Market** 19
5. **After-Effects of Deregulation** 21
6. **Concluding Remarks** 25
   References 25
   Further Reading 26
   List of Related Web-Sites 27

CHAPTER-2: POWER SYSTEM ECONOMIC OPERATION OVERVIEW

1. **Introduction** 29
2. **Economic Load Dispatch (ELD)** 30
   2.1 The Economic Load Dispatch Problem 30
   2.2 Conditions for the Optimum 32
   2.3 A Review of Recent Developments in ELD 34
   2.4 Example 35

3. Optimal Power Flow as a Basic Tool **37**
3.1 The Basic OPF Model 38
3.2 Example 41
3.3 Characteristic Features of OPF 44
4. Unit Commitment (UC) **47**
4.1 UC: The Basic Model 48
4.2 UC: Additional Issues 50
5. Formation of Power Pools **53**
5.1 Power Pools 54
5.2 The Energy Brokerage System 61
6. Concluding Remarks **69**
References 69

CHAPTER-3: POWER SYSTEM OPERATION IN COMPETITIVE ENVIRONMENT

1. Introduction **73**
2. Role of the Independent System Operator (ISO) **74**
2.1 Structure of UK and Nordic Electricity Sector Deregulation 75
3. Operational Planning Activities of ISO **80**
3.1 The ISO in Pool Markets 80
3.2 The ISO in Bilateral Markets 91
4. Operational Planning Activities of a Genco **95**
4.1 The Genco in Pool Markets 95
4.2 The Genco in Bilateral Markets 96
4.3 Market Participation Issues 97
4.4 Unit Commitment in Deregulated Environment 100
4.5 Competitive Bidding 107
5. Concluding Remarks **114**
References 115

CHAPTER-4: TRANSMISSION OPEN ACCESS AND PRICING ISSUES

1. Introduction **119**
1.1 The US and the European Perspective: Transco vis-à-vis the Transmission System Operator 120
2. What is Power Wheeling? **121**
3. Transmission Open Access **121**
3.1 Types of Transmission Services in Open Access 122
4. Cost Components in Transmission **123**
5. Pricing of Power Transactions **125**

5.1 Embedded Cost Based Transmission Pricing 127
5.2 Incremental Cost Based Transmission Pricing 134
6. Transmission Open Access and Pricing Mechanisms in Various Countries 138
6.1 United Kingdom 138
6.2 Chile 138
6.3 Sweden 139
7. Developments in International Transmission Pricing in Europe 142
7.1 Example 143
8. Security Management in Deregulated Environment 145
8.1 Scheduling of Spinning Reserves 147
8.2 Interruptible Load Options for Security Management 148
9. Congestion Management in Deregulation 157
9.1 Economic Instruments for Handling Congestion 158
10. Concluding Remarks 166
References 167

CHAPTER-5: ANCILLARY SERVICES MANAGEMENT
1. What Do We Mean by Ancillary Services? 171
1.1 General Description of Some Ancillary Services 172
2. Ancillary Services Management in Various Countries 175
2.1 The US 175
2.2 UK 178
2.3 Australia 179
2.4 Sweden 182
2.5 Check-List of Ancillary Services Recognized by Various Markets 185
3. Reactive Power as an Ancillary Service 186
3.1 Reactive Power Management in Some Deregulated Electricity Markets: A Review 187
3.2 Defining Scope of the Service is Important 198
3.3 Synchronous Generators as Ancillary Service Providers 200
4. Concluding Remarks 202
References 203

CHAPTER-6: RELIABILITY AND DEREGULATION
1. Terminology 205
2. Reliability Analysis 206
2.1 The Interruption Criterion 208

2.2 Stochastic Components 209
2.3 Component Models 210
2.4 Calculation Methods 216
3. The Network Model **218**
3.1 Stochastic Networks 218
3.2 Series and Parallel Connections 220
3.3 Minimum Cut-Sets 222
4. Reliability Costs **223**
5. Hierarchical Levels **227**
5.1 Generation Reliability 227
5.2 Transmission Reliability 229
5.3 Distribution Reliability 231
6. Reliability and Deregulation **232**
6.1 Is There a Conflict? 232
6.2 Reliability Analysis 236
6.3 Effects on the Actual Reliability 237
6.4 Regulation of the Market 241
7. Performance Indicators **246**
8. Conclusions **248**
References 249
Further Reading 250

CHAPTER-7: POWER QUALITY ISSUES: ***Voltage Dips And Other Disturbances***
1. Power Quality **253**
1.1 Terminology 253
1.2 Interest in Power Quality 255
1.3 Events and Variations 256
2. Voltage Dips **257**
2.1 Magnitude and Duration 257
2.2 Origin of Voltage Dips 259
2.3 Magnitude, Duration and Fault Location 264
2.4 Three-phase Unbalanced Dips 266
2.5 Voltage Dip Performance Indicators 271
2.6 The Responsibility Question 279
2.7 Voltage Dip Mitigation 281
3. Other Power Quality Issues **286**
3.1 Short Interruptions 286
3.2 Harmonic Distortion 288
3.3 Transient Overvoltages 289
3.4 Voltage Variations 290

3.5 Voltage Steps 290
3.6 Voltage Fluctuations 290
3.7 Load Currents 291
4. Conclusions **291**
References 292
Further Reading 293
Appendix A: IEC Standards on Power Quality 294
Appendix B: IEEE Standards on Power Quality 296

INDEX **299**

List of Acronyms

ACE	Area Control Error
ATSOI	Association of Transmission System Operators in Ireland
BTM	Bilateral Transaction Matrix
Cal-ISO	California Independent System Operator
CEGB	Central Electricity Generating Board (UK)
CIL	Contracted Interruptible Load
DED	Dynamic Economic Dispatch
DISCO	Distribution Company
DSB	Demand Side Bidding (UK)
DVR	Dynamic Voltage Restorer
ELBAS	Nordpool Short-term Market
ELD	Economic Load Dispatch
ELSPOT	Nordpool Spot Market
EMC	Electromagnetic compatibility
EPACT	Energy Policy Act (US)
EPRI	Electric Power Research Institute (US)
ERPS	Enhanced Reactive Power Service (UK)
ETSO	European Transmission System Operators
EU	European Union
FERC	Federal Energy Regulating Authority (US)
GAMS	Generalized Algebraic Modeling Systems
GENCO	Generating Company
IEEE	Institute of Electrical and Electronics Engineers
IEC	International Electrotechnical Commission
IPP	Independent Power Producer
ISO	Independent System Operator

ITM	Interruptible Tariff Mechanism
LNG	Liquefied Natural Gas
LPG	Liquefied Petroleum Gas
LOLP	Loss of Load Probability
LRMC	Long-run Marginal Cost
LSE	Load Serving Entity (New York pool)
MC	Marginal Cost
NEMMCO	National Electricity Market Management Company (Australia)
NERC	North American Electric Reliability Council
NETA	New Electricity Trading Arrangements (UK)
NGC	National Grid Company (UK)
Nordel	A body for cooperation between ISOs of Nordic countries
Nordic	Norway, Sweden, Finland, Denmark and Iceland
NUG	Non-Utility Generator
NYISO	New York Independent System Operator (US)
NZEM	New Zealand Electricity Market
OFFER	Office of Electricity Regulation (UK)
OFGEM	Office of Gas and Electricity Markets (UK)
OPF	Optimal Power Flow
ORPS	Obligatory Reactive Power Service (UK)
PGCIL	Power Grid Corporation of India
PJM	Pennsylvania, Jersey, Maryland power pool (US)
PLF	Plant Load Factor
PPP	Pool Purchase Price
PSP	Pool Selling Price
PURPA	Power Utilities Regulating Policy Act (US)
REC	Regional Electricity Company (UK)
RTO	Regional Transmission Organizations (US)
SCED	Security Constrained Economic Dispatch
SMP	System Marginal Price
SPD	Spot Price Difference
SRMC	Short-run Marginal Cost
TRANSCO	Transmission Company
TSO	Transmission System Operator (Europe)
UC	Unit Commitment
UCTE	Union for Coordination of Transmission of Electricity (Europe)
UKTSOA	United Kingdom Transmission System Operators' Association
UPS	Uninterruptible Power Supply
WRATES	Wheeling Rate Evaluation Simulator

Preface

Deregulation is a fairly new paradigm in the electric power industry. And just as in the case of other industries where it has been introduced, the goal of deregulation is to enhance competition and bring consumers new choices and economic benefits. The process has, obviously, necessitated reformulation of established models of power system operation and control activities. Similarly, issues such as system reliability, control, security and power quality in this new environment have come in for scrutiny and debate.

In this book, we attempt to present a comprehensive overview of the deregulation process that has developed till now, focussing on the operation aspects. As of now, restructured electricity markets have been established in various degrees and forms in many countries. This book comes at a time when the deregulation process is poised to undergo further rapid advancements.

It is envisaged that the reader will benefit by way of an enhanced understanding of power system operations in the conventional vertically integrated environment vis-à-vis the deregulated environment. The book is aimed at a wide range of audience- electric utility personnel involved in scheduling, dispatch, grid operations and related activities, personnel involved in energy trading businesses and electricity markets, institutions involved in energy sector financing. Power engineers, energy economists, researchers in utilities and universities should find the treatment of mathematical models as well as emphasis on recent research work helpful. The book may be used for a one-semester graduate or under-graduate course, as well.

OUTLINE OF THE BOOK

Chapter-1 discusses the motivating factors behind deregulation of the power sector and the after-effects of the same. In this context it looks at issues specific to developed and developing nations.

Chapter-2 focuses on the established models of operational planning activities such as economic load dispatch, unit commitment and optimal power flow. Topics such as inter-utility power transactions, power pools, power wheeling and energy brokerage systems have been discussed. This chapter lays the foundation for the discussions in the subsequent chapters.

Chapter-3 analyzes different market models, and the operational planning issues specific to these, from the perspective of, both, the independent generator and the system operator. The requisite model development and reformulation demanded by a competitive environment is then discussed.

Chapter-4 identifies transmission management issues and then discusses the mechanisms by which these are addressed in the various forms of deregulated structures. The issues covered here are pricing, security and congestion management.

Chapter-5 focuses on ancillary service management, their categorization, and pricing mechanisms as practiced in different electricity markets. Reactive power management in deregulated markets has been examined in detail.

Chapter-6 on reliability and deregulation treats the basics of reliability analysis of power systems. Both system reliability and reliability experienced by the customer is discussed. The chapter also presents ways of characterizing the observed reliability of supply. The role of the office of electricity regulation (OFFER) in the UK system has been discussed.

Chapter-7 presents an overview of the various power quality issues. Emphasis is on issues strongly related to both the utility and the customer such as short interruptions, voltage dips and harmonics. For each of these phenomena an overview of available analysis techniques is given. The possible effect of deregulation on these aspects of power quality is discussed.

ACKNOWLEDGEMENTS

We are grateful to Professor M. A. Pai of the University of Illinois at Urbana-Champaign, USA, for proposing this monograph and then providing us with his invaluable guidance at every stage.

The material used in this book has been taught at graduate level courses at the Chalmers University of Technology as well as for short industrial courses.

Several research projects related to the material covered in this book were generously sponsored by the Swedish power industries and we are indebted to them. In particular, we are grateful to ABB Automation Products, ABB Corporate Research, ABB Power Systems, Elektra Program of Elforsk, Energimyndigheten, Göteborg Energi, Svenska Kraftnät, Sydkraft Research Foundation and Vattenfall.

Our thanks are also due to many persons from the Swedish power industries for their participation and collaboration in our research efforts. In particular, we would like to thank Dr. Daniel Karlsson and Dr. Murari Saha of ABB Automation Products, Mr. Sture Larsson and Mr. Klas Roudén of Svenska Kraftnät, Mr. Lars Sjunnesson, Mr. Alf Larsen and Mr. Gunnar Ridell of Sydkraft and Mr. Bernt Hansson of SWECO Energuide.

We would also like to thank Mr. Alastair Ferguson (Scottish Power), Mr. Rober Olofsson (Unipower), Mr. Christian Roxenius (Göteborg Energi), Mr. Helge Seljeseth (SINTEF Energy Research) and Mr. Mats Häger (Swedish Transmission Research Institute) for supplying us with the measurement data required for the work on power quality.

The Department of Electric Power Engineering at Chalmers University of Technology has always provided a very congenial atmosphere to carry out our work and we are grateful to all the staff of the department. Ph.D. students Adrian Andreou, Mattias Jonsson, Johan Lundquist, Manos Styvaktakis, Le Anh Tuan, Lidong Zhang and Jin Zhong provided critical inputs for various chapters and we gratefully acknowledge their contributions.

We are thankful for the cooperation and help received from the Publishing Department of Kluwer Academic Publishers, in particular Mr. Alex Greene.

Finally, we are very grateful to our families - our wives Rajendrani, Irene and Inga-Britt - for bearing with us all the while and for their support and patience with us. Rajendrani also helped in editing and proofreading the manuscript.

Gothenburg, Sweden
March 21, 2001

Kankar Bhattacharya
Math Bollen
Jaap Daalder

Chapter 1

Deregulation of the Electricity Supply Industry

1. INTRODUCTION

During the nineties decade, many electric utilities and power network companies world-wide have been forced to change their ways of doing business, from vertically integrated mechanisms to open market systems. The reasons have been many and have differed over regions and countries.

Among the developing countries, the main issues have been a high demand growth coupled with inefficient system management and irrational tariff policies, among others. This has affected the availability of financial resources to support investments in augmenting generation and transmission capacities. In such circumstances, many utilities were forced to restructure their power sectors under pressure from international funding agencies. In developed countries, on the other hand, the driving force has been to provide customers with electricity at lower prices and to offer them a greater choice in purchasing economic energy.

Reforms have been undertaken by introducing commercial incentives in generation, transmission, distribution and retailing of electricity, with, in many cases, large resultant efficiency gains. The electricity bill now reflects at least two components: one from the distribution and transmission network operator responsible for the network and services, and the other from the company that generates the electrical energy.

Though this may seem to be fairly straightforward at first glance, there are several complexities involved in restructuring and many new issues have surfaced. Some of them have been solved while others are being discussed at various levels.

2. WHAT IS DEREGULATION?

The electric power industry has over the years been dominated by large utilities that had an overall authority over all activities in generation, transmission and distribution of power within its domain of operation. Such utilities have often been referred to as *vertically integrated utilities*. Such utilities served as the only electricity provider in the region and were obliged to provide electricity to everyone in the region.

The utilities being vertically integrated, it was often difficult to segregate the costs incurred in generation, transmission or distribution. Therefore, the utilities often charged their customers an average tariff rate depending on their aggregated cost during a period. The price setting was done by an external regulatory agency and often involved considerations other than economics.

Figure 1 shows the typical structure of a vertically integrated utility where links of information flow existed only between the generators and the transmission system. Similarly, money (cash) flow was unidirectional, from the consumer to the electric utility.

The operation and control issues for such systems have been widely examined over the years. The basic objective of the operator in such vertically integrated utilities would be to minimize the total system cost while satisfying all associated system constraints. Apart from operational issues, such vertically integrated utilities also had a centralized system of planning for the long-term. All activities such as long-term generation and transmission expansion planning, medium term planning activities such as maintenance, production and fuel scheduling were coordinated centrally.

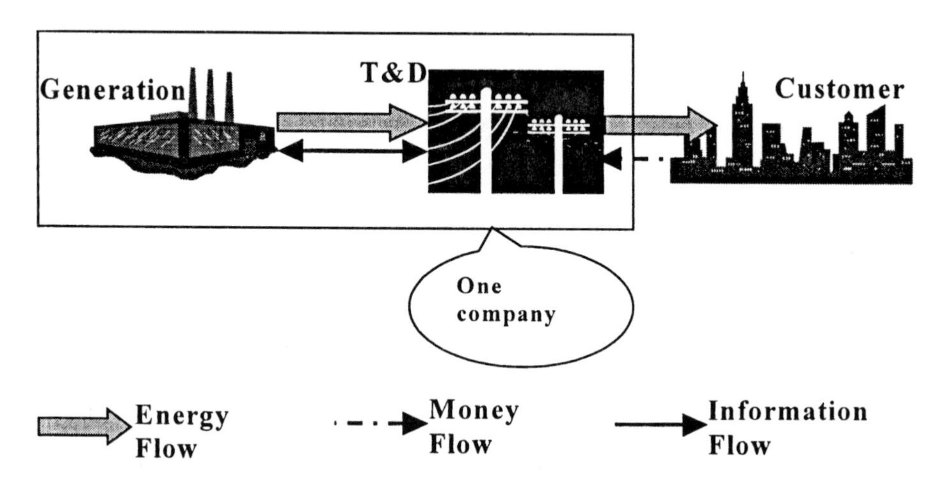

Figure 1. Typical structure of a vertically integrated electric utility

In spite of the fact that such operations, control and planning arrangements seemed to work perfectly well over the years, the electric power industry has been undergoing a process of transition and restructuring since the nineties decade or so.

One of the first steps in the restructuring process of the power industry has been the separation of the transmission activities from the electricity generation activities. The subsequent step was to introduce competition in generation activities, either through the creation of power pools, provision for direct bilateral transactions or bidding in the spot markets.

On the other hand, the transmission system having significant *economics of scale* consequently had a tendency to become a *monopoly*. Thus it was felt necessary to introduce regulation in transmission so as to prevent it from overcharging for its services. The transmission grid thus became a *neutral, natural monopoly* subject to regulation by public authorities. And to overcome the monopolistic characteristic, the trend has been to establish new legal and regulatory frameworks offering third parties *open access* to the transmission network.

An important point to note is that the restructuring process was however not uniform in all countries. While in many instances, it started with the breaking up of a large vertically integrated utility, in certain other instances restructuring was characterized by the opening up of small municipal monopolies to competition.

A system operator was appointed for the whole system and it was entrusted with the responsibility of keeping the system in balance, *i.e.* to ensure that the production and imports continuously matched the consumption and exports. Naturally, it was required to be an independent authority without any involvement in the market competition nor could it own generation facilities for business (except for owning some capacity for emergency use). Quite appropriately, the system operator came to be known as the Independent System Operator (ISO).

Figure 2 shows the typical structure of a deregulated electricity system with links of information and money (cash) flow between various players. Note that this, again, is not a universal configuration and that there exists variations across countries and systems. The possibility of having such a complex nature of information flow has been one of the driving factors in the process of deregulation of the power sector. This has been possible due to the rapid developments in the fields of communication and information technology during the nineties decade.

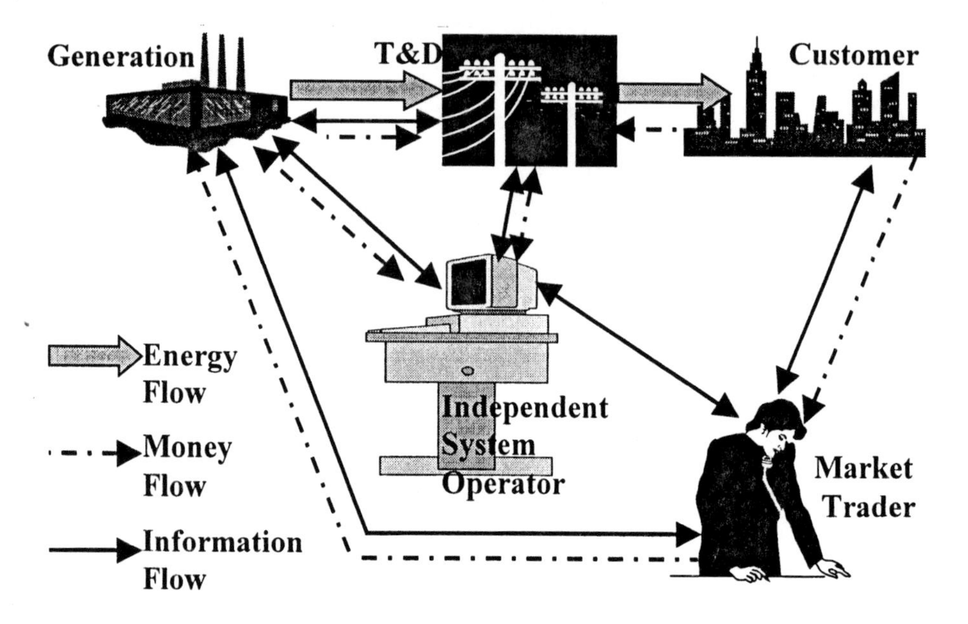

Figure 2. Typical structure of a deregulated electricity system

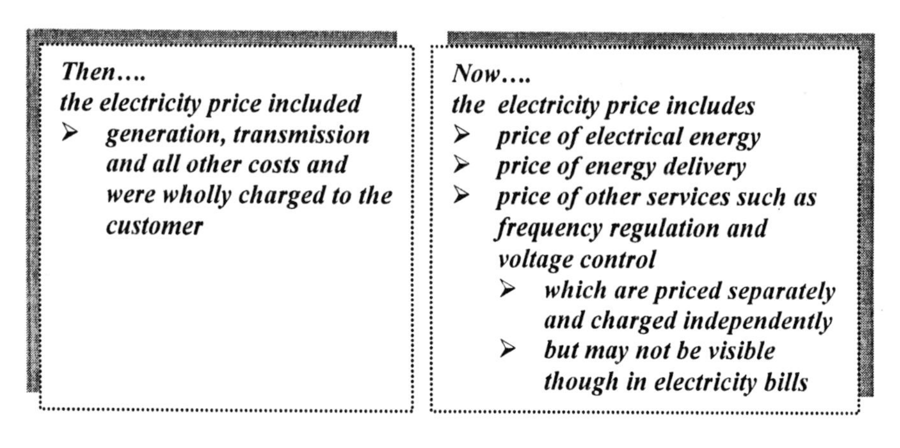

Figure 3. Then and Now

2.1 Different Entities in Deregulated Electricity Markets

The introduction of deregulation has brought several new entities in the electricity market place, while on the other hand redefining the scope of activities of many of the existing players. Variations exist across market structures over how each entity is particularly defined and over what role it plays in the system. However, on a broad level the following entities can be identified:

2.1.1 Generator companies (also referred to as *gencos*)

The generators produce and sell electricity. This may refer either to individual generating units or more often to a group of generating units within a single company ownership structure with the sole objective of producing power, and commonly referred to as *independent power producers (IPP).* Different markets may classify generators based on their rated capacity or in the way the generators have been contracted to operate in the market.

2.1.2 Transmission Companies (also referred to as *transcos*)

The transmission companies are those entities, which own and operate the transmission wires. Their prime responsibility is to transport the electricity from the generators to the customer, and making available the transmission wires to all entities in the system. For their services, they levy a transmission tariff. In some systems, these transcos are classified according to the operating voltage levels, such as national transcos (at 400 KV and 220 kV), regional transcos (at 132 kV), *etc.*

2.1.3 Distribution Companies (also referred to as *discos*)

The distribution companies are usually those entities owning and operating the local distribution network in an area. They buy wholesale electricity either through the spot-markets or through direct contracts with gencos and supply electricity to the end-use customers.

2.1.4 Customers

A customer is an entity consuming electricity. In deregulated markets, the customer has several options for buying electricity. It may choose to buy electricity from the spot-market by bidding for purchase, or may buy directly from a genco or even from the local distribution company.

2.1.5 Independent System Operator (ISO)

The ISO is an entity entrusted with the responsibility of ensuring the reliability and security of the entire system. It is an independent authority and does not participate in the electricity market trades. It usually does not own generating resources, except for some reserve capacity in certain cases. In order to maintain the system security and reliability, the ISO procures various services such as supply of emergency reserves, or reactive power from other entities in the system.

2.1.6 Market Operator

The Market Operator is an entity responsible for the operation of the electricity market trading. It receives bid offers from market participants and determines the market price based on certain criteria in accordance with the market structure. The markets may have different trading schemes such as hourly trading for the next day or trading in futures- weeks, months or years ahead.

3. BACKGROUND TO DEREGULATION AND THE CURRENT SITUATION AROUND THE WORLD

In the previous section we introduced a basic understanding of *deregulation*. We mentioned that during the nineties decade there has been a fundamental change in the thinking about the structure of electricity supply industries around the world. These changes have largely followed a common pattern- competition in power generation facilitated by an independent transmission network. But within this framework, there has been considerable diversity and various structures of deregulation have come into existence in different countries.

The basic question that arises at the outset is- *why deregulation?* We shall attempt to address this question by examining the background issues that led to deregulation. In this context two different perspectives are brought out, *i.e.* those issues relevant to the industrialized countries and those relevant to the developing ones.

3.1 Industrialized Countries

The industrialized countries have, over the years, had a well functioning and, often, quite efficient electricity systems in place when their deregulation processes started. We shall examine some specific country cases and reasons for their deregulation. A common factor that was more or less prevalent in all these countries was that either the customers were not satisfied by the rising costs of electricity or in some cases the utility management found the operations not viable due to low tariffs. In certain other cases deregulation was the result of pressure from the smaller players in the business to reduce the control and power of large state-owned large utilities by opening up the market to competition.

Thus deregulation precipitated in the industrialized countries by the pressure to reduce costs and hence tariffs while, simultaneously increasing the competitiveness in the markets.

3.1.1 In the US

While in most of the European deregulation processes, there was a stage of unbundling of utilities, followed by a stage of privatization; the US electric utilities were, from the very beginning predominantly investor-owned (*i.e.* private). This structure therefore required a different form of restructuring.

The Public Utilities Regulatory Policy Act (PURPA) of 1978 effectively initiated the deregulation process in US by allowing non-utility generators (NUG) to enter the wholesale power market. The Act required the utilities to purchase surplus electricity from the NUG at prices up to its avoided cost. This policy incentive allowed for a substantial growth in NUG generating capacity during the nineties decade. From a total NUG generating capacity of 42,000 MW in 1989, the capacity increased to 98,000 MW by 1998 accounting for a growth rate of more than 200%. Accordingly, the energy generation from NUG increased rapidly and by the end of 1998, accounted for more than eleven per cent of total US electricity supply [1]. *Figure 4* shows the year-wise growth of generation from the NUG and their share in total US energy supply.

Subsequent to PURPA, the US Energy Policy Act of 1992 (EPACT) provided the thrust for the development of competitive power markets in the US. It mandated the electric utility industry to become deregulated and ordered the Federal Energy Regulatory Authority (FERC) to facilitate this transition.

In April 1996, FERC issued final rules on Open Transmission Access (Orders 888 and 889), thereby facilitating the transition and requiring the transcos to allow non-affiliated selling / buying organizations to access their transmission systems in a non-discriminatory manner. The primary aim of these Orders was to eliminate monopoly power over the transmission of electricity [2, 3].

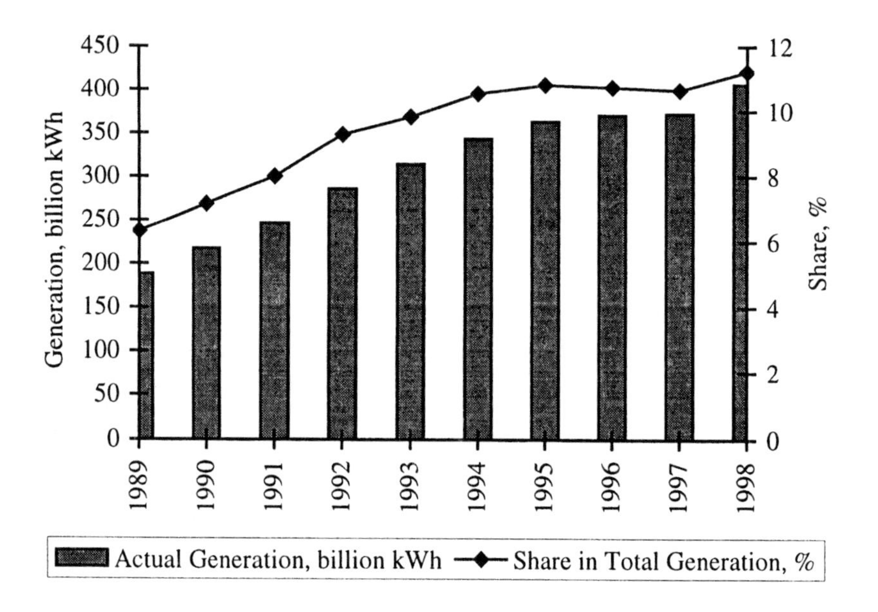

Figure 4. Growth in NUG participation in US electricity sector during the Nineties. *Source: Energy Information Administration,* http://www.eia.doe.gov

More specifically, it required all public transcos to provide open access transmission services with non-discriminating tariffs, develop and maintain a *same-time information system* that provides all users the same access to transmission information that the transco enjoys.

In December 1999, FERC issued Order 2000 [4, 5], with an objective to reform the operational practices of transcos. Under this, the transcos would be required (though voluntarily) to re-organize themselves into different Regional Transmission Organizations (RTO) in order to address the operational and reliability issues in transmission and eliminate any residual discrimination in transmission services. The RTO is envisaged to undertake the sole responsibility for operation and expansion of the transmission system as well as for transmission tariff settings. Although the RTO can in principle, be a profit making entity, its scope of activities vis-à-vis that of the ISO is still being discussed and debated.

These initiatives of the US electricity authorities broke up the traditional vertically integrated utility model into various organizations composed of generation, transmission and distribution entities and required them to operate independent of each other. Many new entities such as the independent power marketers, independent power producers, independent system operators and power exchanges were born.

3.1.2 The Scene in Europe

The European Union (EU) Directive on the Internal Electricity Market came into force on 19th February 1999 [6]. With this directive it was envisaged, over a period of time, to allow all large and medium sized purchasers of electricity to choose their suppliers freely from throughout the EU. The EU Directive introduces full competition amongst generators immediately and is already leading to significant price reductions across the EU to the benefit of business and consumers (*Figure 5*).

The directive however, does not impose a single rigid new market structure for all countries, but sets out the minimum conditions under which competition can develop in a fair and transparent way.

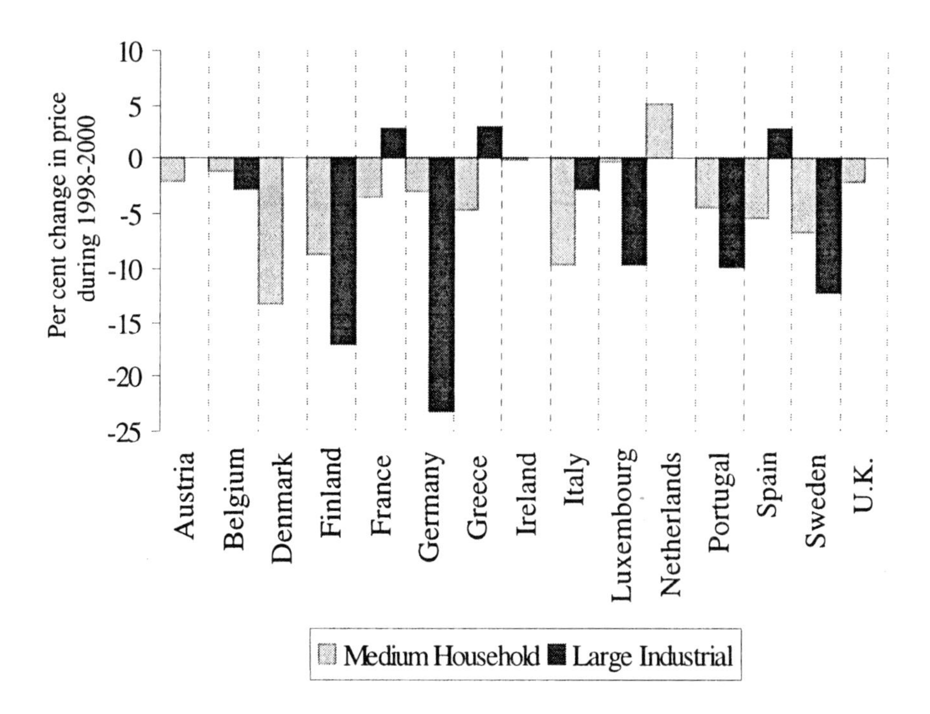

Figure 5. Per cent change in electricity prices (excluding tax) for medium household consumers (annual consumption: 3500 kWh, subscribed demand: 4-9 kW) and large industrial consumers (annual consumption: 50 GWh, Maximum demand: 10 MW, annual utilization: 5000 hours) in various EU countries during 1998-2000. Information on industrial sector is not available for Austria, Denmark, Netherlands and UK while for Ireland, there was no change in price. *Source: Eurostat,* http://europa.eu.int/comm/eurostat/

As a result of the opening of the electricity market, cross-border trading is rapidly increasing. On 1st July 1999, a European body exclusively for transmission system operators was founded, The Association of European Transmission System Operators (ETSO). The principle objective of ETSO is to promote the conditions for an efficient European electricity market. One of the tasks of the organization is to help formulate an effective regulatory framework for transmission of power between countries, and effective price structures for power transmission. ETSO, which covers all the 15 EU member states, and in addition Norway and Switzerland, has been formed out of four existing transmission system operators' associations: Nordel (in Nordic[1] countries), ATSOI[2] (Ireland), UCTE[3] (western continental Europe) and UKTSOA[4].

Subsequent developments suggest that the Nordic and the British model of ISO with overall system wide responsibility will be the favored option in most countries and of the European Commission. Such an integrated company will be designated as the Transmission System Operator (TSO) by the regulatory authorities.

Among some of the other specific salient features of the Directive are:

- Investment in power generation and construction of new power plants can henceforth be made anywhere within the EU subject to meeting the set procedures of member countries.

- Large and medium-sized customers will be able to choose their electricity supplier. A step by step approach, beginning with very large consumers in 1999 and subsequently covering all medium sized customers by 2003, has been proposed. However, most of the countries are moving at much faster rates than this stipulated rate.

- The transmission and distribution network owners are required to provide access to their lines to others. Three alternative models have been proposed- *regulated third party access*, *negotiated third party access*, and *the single buyer model.* While twelve countries have opted for regulated third party access, the remaining three have opted for negotiated third party access.

3.1.2.1 The British Power Pool

Restructuring of the UK power sector was set in motion in 1988 which subsequently led to the breaking up of the erstwhile Central Electricity

[1] Norway, Sweden, Finland, Denmark and Iceland are known as the *Nordic* countries
[2] Association of Transmission System Operators in Ireland
[3] Union for Coordination of Transmission of Electricity
[4] United Kingdom Transmission System Operators' Association

Generating Board (CEGB) which owned nearly 60,000 MW of capacity and all the high voltage transmission lines in England and Wales. Four companies, namely, National Power, Power Gen, Nuclear Electric and National Grid Company (NGC) emerged from the split, the latter two being state-owned companies. NGC was made responsible for the national transmission system while there were 12 Regional Electricity Companies (REC) managing the sub-transmission and distribution networks. The NGC also assumed the role of the system operator with the responsibility of promoting competition [7].

The NGC manages the activities of the market by way of organizing the bidding process, receiving bids, settling the market prices every day including unit commitment and dispatch. It is also responsible for transmission system security, reliability, congestion management and setting up transmission prices. Participation of all the generators in the power pool is *mandatory*. The total load forecast for the system is available to NGC since all RECs submit their regional forecasts[5]. The pool model is discussed in more detail in Chapter-3.

3.1.2.2 Nordic Deregulation Process- Bilateral Contracts Dominate

3.1.2.2.1 Sweden

The Swedish electricity sector was never completely centralized or nationalized. Till 1991, the sector was dominated by *Vattenfall*, which in addition to owning about 50% of the total generation also managed the 400 kV and 220kV transmission lines, and some large networks at lower voltage levels, down to the customers. There were about a dozen other large generating companies and 270 distribution companies, which operated the networks at lower voltage levels and often owned their own generation.

The large generating companies had a joint agreement- the *Pooling Agreement* aimed at optimizing the operation of the generating resources. This arrangement was similar to a pool, but it must be stressed here that there was no centralized dispatch.

[5] The independent electricity regulating authority in UK, the Office of Gas and Electricity Markets (OFGEM) has recently set into motion discussions on New Electricity Trading Arrangements (NETA) and is expected to replace the current pooling arrangements by spring 2001.

NETA has proposed that forward and future markets will evolve on bilateral contracts basis. Short-term power exchange will be available for participants to fine tune their requirements and the system operator, NGC, will manage a balancing mechanism based on bid offers. More recent information will be available on OFGEM web site:

http://www.ofgem.gov.uk

The basic requirement was that each power company owned enough generating capacity to be able to meet its dedicated customer demand during normal operating conditions. As has been the established norm in the Nordic countries, most of the generating companies used to enter in to *direct bilateral contracts* with the customers.

Each company planned and dispatched its own generation, but the running costs of any spare available generating resources were continuously made known to the other members of the pool. Trade was mostly carried out during the day of delivery on an hourly basis, but there was also a small amount of weekly and monthly trading.

The price was calculated by splitting the difference between the buyer's and the seller's declared marginal operation costs at the geographical point of interchange. To be able to do this, the companies had to have access to the transmission networks to convey the purchased power to their customers.

The natural central player in this set up was, again, Vattenfall. Most of the other generating companies ultimately formed a joint organization where they first pooled their generating resources before meeting Vattenfall. The other larger private generating companies saw the possible advantages to Vattenfall from such an arrangement. Also, the smaller players did not have any access to the market for supplying power occasionally [8].

In 1991, the Swedish government decided to remove transmission activities from Vattenfall and create the state owned transmission company, Svenska Kraftnät, to manage the national transmission network. In addition, Svenska Kraftnät was assigned the task of promoting competition in the electricity market. One aim of this reorganization was to open up the national transmission network and interconnections to companies other than the large generating companies.

From January 1, 1995, Svenska Kraftnät introduced its new transmission tariff based on *point of connection*, which aimed at promoting competition on the electricity market. The basic principle was that with a single payment, at the point of connection, a customer could access the entire network system and thus get into trading arrangements with any player on the system[6].

A few large producers account for more than 90% of the total electricity generation in Sweden and more than 220 supply companies compete in the market. From 1st November 1999 the requirement for specific electricity meters or other extra charges for ordinary customers have been abolished in order to make it easier for customers to choose a new supplier freely.

[6] This scheme comes under the broader classification of *regulated third party access* model of transmission open access, mentioned earlier.

3.1.2.2.2 Norway

The Norwegian electricity sector was dominated by small / medium sized municipality owned power companies, each vertically integrated, i.e., they generated power and transmitted that to their own dedicated customers. Most of the transactions were on a bilateral basis, between the utility and the bulk customer. The system being completely (100%) hydro based, there was enough volatility of prices, depending upon water availability and therefore a regular mechanism of power import from Sweden was in place. Statkraft was the largest player in this market, owning about 35% of the generation, in addition to the transmission network.

In 1991, Statkraft was divested of its transmission responsibilities and these were passed on to the newly established Statnett, which was also assigned the role of the system operator. Statkraft remained a major generating company.

A tariff system was introduced (grid access tariff)- a prerequisite for the customers' ability to choose their suppliers freely.

Presently, approximately 200 utilities are competing to supply electricity to Norwegian customers who can choose a supplier freely and at no charge. There are no requirements as to specific electricity meters for small customers.

3.1.2.2.3 NordPool: The Nordic Electricity Market

As a subsidiary of Statnett, the Statnett Power Market was created in 1993 to operate the bulk power market in Norway. In 1994, Statnett and Svenska Kraftnät started investigating into the possibility of a joint Norwegian-Swedish electricity trading exchange. The arguments for a common trading exchange were that Norway's electricity market had already been reformed and Sweden was on the verge of a similar reform. The countries had many interconnections with a maximum total capacity of 2700 MW. Both had a network company responsible for transmission and system balance and both had introduced the point-of connection tariff and regulating market.

Subsequently NordPool was established in 1996, the first international power exchange in the world, owned equally by Statnett and Svenska Kraftnät. Subsequently, Finland joined the NordPool power exchange in 1998 and Denmark in 1999[7].

NordPool operates two types of markets, a futures market and a spot market. In the futures market, contracts for purchase of base or peak load on a weekly basis for up to three years in advance are possible.

[7] The Danish power network consists of an eastern grid synchronously connected to Sweden, and a western part synchronously connected to Germany. In the NordPool spot market, these two areas therefore have separate market settlements and prices.

3.1.3 Australia and New Zealand

The Australian electricity supply industry also operated as a largely vertically integrated system till 1990. A restructuring of the utilities was initiated in 1991 in order to enhance the overall system efficiency. The Industry Commission, in a background study, reported that the Australian electricity and gas sector reforms could increase the country's gross domestic product by 1.4 per cent a year in 1993-94 dollar terms. By 1994, the first wholesale electricity market started operating in the state of Victoria and by 1998 a full-fledged National Electricity Market (NEM) came into being. The National Electricity Market Management Company (NEMMCO) was entrusted with the role of the ISO as well as of the market operator [9].

The Australian market structure is similar to the UK power pool and NEMMCO receives bids from generators and wholesale market customers for selling / purchase of power on a half-hourly basis. A centrally coordinated dispatch is carried out by NEMMCO using the bid price offers from participants and a demand forecast for the said half-hour, to determine the market-clearing price, that is then applicable to all participants. The generators' bids provide information on (a) energy price offer (b) power offered for sell and (c) unit availability. The customers bid for energy purchase as well as interruptible loads. All market participants are provided with information on demand and generation forecasts, planned line outages, generator availability as well as exports / imports between regions.

There are three types of direct energy trading: a) spot market trades accounting for about 10% of the volume b) vesting contracts, accounting for about 35% and c) bilateral contracts about 60% [10]. In addition to the *Direct Market Trading*, there exists a secondary hedge market, which helps in reducing the market risk through a futures market or through hedges between power stations, which can contract for capacity interchanges to take care of contingency situations

The New Zealand Electricity Market (NZEM) began its operation on 1st October 1996. The trading is conducted by the Electricity Market Company (EMCO) based on a day-ahead settlement from bids submitted at half-hour blocks [11]. The bids are specified by commodity type (energy bid or reserve bid). This means that the suppliers have to simultaneously bid for energy supply and / or reserves in to the market. However, spot market participation is not mandatory. Another interesting feature of this spot-market is that the bidders can revise their bid offers up to four hours prior to real time and the market is re-cleared on a regular basis. Commodity type, time, as well as location, in addition to their quantity and price offers, therefore specify a bid. Energy and reserves are cleared simultaneously by the system operator.

Prices are calculated for 48 half-hour trading periods every day. The demand bids, arranged in decreasing order of prices are matched with the energy supply bids, arranged in the increasing order. The price varies across nodes according to the transmission losses and grid constraints. This enables more effective risk management on the part of market participants and also sends strong investment signals to generation and transmission owners.

In New Zealand, Trans Power has the responsibility of system operation and is responsible for the national grid. It co-ordinates grid operations. It is also responsible for dispatch and scheduling of generation taking into account the security of the system.

Further, Trans Power is also responsible for procuring ancillary services from ancillary service providers. On the whole, its role in the system management is similar to that played by NGC in the UK system.

3.2 Developing Countries

In most of the developing countries, the growth of the power sector was under the direct supervision of their respective federal governments. They held the sole responsibility on all fronts such as in making investment decisions for new power projects, providing for budgetary support within their annual plan outlays, setting the operating guidelines for their generators and transmission systems, and finally setting the prices for the customers.

Electricity supply was treated more as a social service, than a marketable commodity, as an essential input to the building up of a sound infrastructure. However, such state-controlled system management led to the promotion of inefficient practices- both technical and managerial. Some of these inefficiencies are discussed below [12]:

➢ **Inefficiency in Production, Transmission, Distribution and Use**

This includes inefficiency within the generating stations in terms of high auxiliary consumption (sometimes to the order of 14% of the generation), specific oil consumption and specific coal consumption. Various factors were responsible for this- external factors such as poor coal quality; grid specific factors such as poor plant load factor[8] (PLF) operation and high reactive power generation; plant specific factors such as operational constraints; and managerial factors relating to the overall plant practices observed.

PLF has a bearing on the auxiliary consumption and hence generation efficiency. There may be several reasons for a low PLF such as backing down

[8] Plant load factor of a generating unit or station is defined as the ratio of actual energy generated by it during a day, week, month or year to its maximum energy generation capability during the same period, expressed in per cent.

of units due to demand constraints, outages due to poor quality coal, operational problems resulting in reduced availability of the plant, *etc.*

Inefficiencies in transmission and distribution relate to the losses incurred, both *technical* and *commercial*, in transporting the electricity from the generating station to the consumer. The high *technical losses* may be due to increased transformer losses from overloads and poor maintenance. It may also include feeder losses due to low load factors, and low and uneven demand that are typical in rural areas. In urban areas, overloaded feeders, low power factor loads, low transmission voltages, long feeder lengths, and many other factors account for high losses.

Commercial or Non-technical losses comprise the losses arising from pilferage of energy, inaccurate meter readings, defective meters, *etc.* This is quite a significant component in many electric utilities in developing countries.

➢ **Irrational Pricing Policies**

An important source of inefficiency in electric utilities has been the existence of irrational tariff policies. Most often the utilities have no commercial autonomy (managerial or financial). The governments often implemented social subsidy policies through the utilities thereby compromising the financial viability of the latter. Most of them subsidize the agricultural consumers who are not even charged the marginal cost of generation. In practice, the electric utilities find vital commercial subjects such as tariffs, new connections and discontinuance of connections beyond their purview. This has often kept them from operating on a fully commercial basis.

Figure 6 shows the average tariff charged from various consumer categories in India during 1995-96. It is evident that the domestic and agricultural consumers were heavily subsidized in the tariff rates while the industrial and commercial sectors paid more. It is also important to note that, in spite of the cross-subsidization, the tariff policy failed to recover the total costs incurred by the system (note the average cost and average tariff rates).

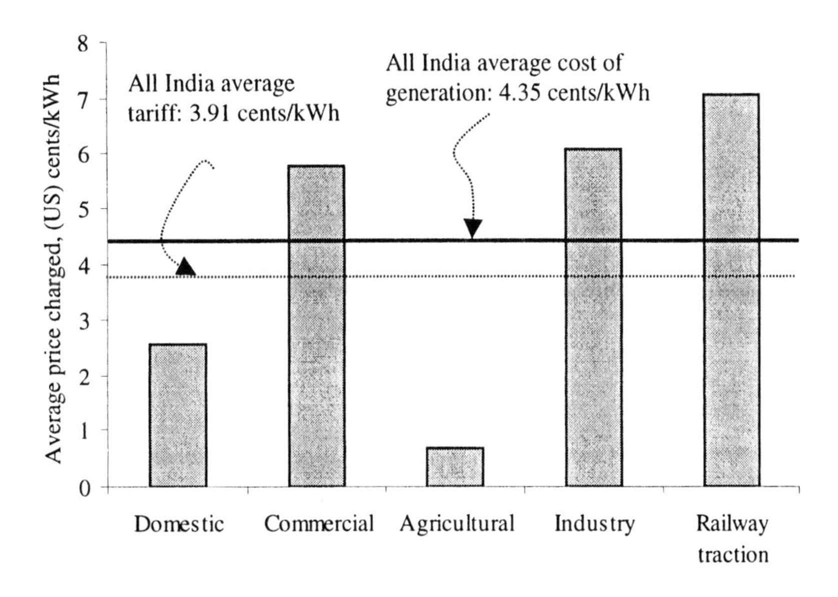

Figure 6. Average electricity tariff charged from different consumer categories in India (1995-96). *Source: Planning Commission Reports on State Electricity Boards of India*

- **Overstaffing in the Utilities**

Yet another consequence of state-control over electric utilities was the enormous level of overstaffing. An electric utility with geographically widespread distribution system provided an ideal opportunity to offer jobs either to fulfil election promises or to dispense favors. Apart from heavy financial costs, overstaffing resulted in inefficiency, lack of unaccountability and mismanagement of resources.

These have been some of the common factors prevalent in most developing countries, added to the fact of having a high demand growth and the need for additional investments of the order of billions of dollars. Naturally, the state-owned utilities have been unable to provide with the necessary funding for such capacity-addition requirements thereby inviting financial support from international financial institutions such as the World Bank and the International Monetary Fund.

Most of the funding that was made available from such institutions came with conditions of opening up the electricity sector. The system of *built, own, operate and transfer* strategy was introduced and that involved project construction, generation of electricity for sale to the state utility and the eventual transfer of the plant to the host country after a period, usually between ten and twenty-five years [13].

3.2.1 South America: The Initiators

The countries in the South American continent were faced with problems common to developing countries, *i.e.* the lack of funds for catering to investment requirements. This could mainly be attributed to sub-optimal pricing strategies and other forms of mismanagement in the system. Consequently, the funding grants from the international agencies were conditional upon restructuring of their electricity sectors.

The path breaking restructuring of Chile's electricity sector in 1982 induced several other countries in the Latin American region such as Argentina, Bolivia, Colombia, Peru and Brazil to introduce reforms. The impact of the reforms was soon visible as losses in the transmission and distribution networks in Chile came down from 21% in 1986 to 8.6% in 1996. The electricity prices in Argentina reduced from 39$/MWh in 1994 to less than 32$/MWh in 1996. Productivity in terms of energy generated per employee almost doubled in Chile's largest electricity company, Endesa, during the period 1989 to 1996 [14, 15]; and so on.

The outcome of restructuring was very positive, the electricity sector efficiency has increased many folds and competition has increased with the formation of many competing generators and the countries as a whole consequently attracted more investments from the developed countries.

All the countries opted for the centralized pool type of model where generation scheduling is centralized and have adopted a nodal pricing scheme for the wholesale trades.

3.2.2 In Transition: India

The installed capacity of publicly owned generation in India was about 101,000 MW as of January 2001 and this excludes a substantial generation capacity existing in the NUG sector. In spite of this, the Indian power sector faces an endemic shortage- an energy shortage of about 8% and a peak load shortage of about 10% on the average. This can be attributed to the high demand growth rate of about 8% per year, which has been persisting since the nineties decade.

It is estimated that with a GDP growth rate of 6% per year, the country will require an additional 70,000 MW during the period 2000-2007. At a very approximate cost estimate of $1000/MW for generating capacity and $700/MW for supporting transmission and distribution facilities, the gross investment required, is an overwhelming amount of $140 to 200 billion [16].

Till 1991, around 90% of the investments came from the public sector through Five Year/ Annual Plans. However, from the Eighth Plan onwards (1992-97), it was realized that the public sector would not be able to come up with the finances needed for additional generation capacity requirements, to the tune of $200 billion [12].

An important development thereupon was the government decision in 1991 to encourage private investment, both domestic and foreign, in generation. With the federal and state governments less willing and / or less able to pay for capacity investments, there has been great interest in private investors, both domestic and foreign, to enter the Indian power market. The reason that foreign investors have been so far leading this investment drive is obvious: huge growth potential.

3.2.2.1 Private Participation in Transmission

After the recent amendment in Electricity Laws, transmission has been accorded an independent status and the concept of central and state transmission utilities been introduced. While Power Grid Corporation of India (PGCIL) has been notified as the Central (federal) Transmission Utility (CTU), the state transmission companies would be the state transmission utilities, which are mandated to be government companies.

The participation of the private sector in transmission is proposed to be limited to the construction and maintenance of transmission lines for operation under the supervision and control of the federal or state transmission utility. The process would involve identification of transmission lines by the transmission utility to be entrusted to the private sector, issuing of specifications, inviting offers / bids, and selection of a private party. The transmission utility would then recommend to the federal or state electricity regulatory commission for issuance of a transmission license to the selected private company. The private company shall contract only with the appropriate transmission utility for the entire use of the transmission lines constructed by the company, and shall be responsible for the maintenance of the lines. Transmission charges payable to the company would be directly linked to the availability of the lines.

4. BENEFITS FROM A COMPETITIVE ELECTRICITY MARKET

Opening up the electricity sector to competition is an important tool to improve the efficiency of power generation and thereby to benefit consumers. The vertically integrated utilities could recover their costs regardless of whether they operated efficiently or not. However, with the introduction of competition, there has been an important shift from this approach. Producers have ceased to be protected by their exclusive rights to generate and supply electricity. Competitive markets provide the driving force for generators to innovate and operate in the most efficient and economic manner in order to remain in the business and recover their costs.

The competitive electricity market thereby offers customers and industry participants a range of benefits. Most of the benefits accrue from the downward pressure on electricity prices as industry participants compete to secure purchase of their electricity and services. Other benefits ensuing from the introduction of a competitive electricity market include the following:

- Cheaper electricity: Cheap electrical energy increases the attractiveness of a region as a site for new industry and business opportunities. Lowering production costs for energy intensive customers will allow them to re-invest more profits back into their businesses.

- Efficient capacity expansion planning: Investment decisions are enhanced through greater knowledge of the demand-supply dynamics, allowing generating companies to invest at appropriate locations and time. New participants are encouraged to enter the various competitive sectors of the power industry, thereby enhancing economic development. These are the Independent Power Producers (IPP).

- Pricing is cost reflective, rather than a set tariff: Price signals drive the competitive market, encouraging industry participants to minimize the cost of supplying electricity to customers.

- Cost minimization: Results from the pressure on industry participants created by increased competition in selective sectors, third party access arrangements and independent economic regulation of the natural monopoly functions.

- More choice: Customers have more choice in a competitive market, as retailers vie for their business by offering a range of options for buying electricity.

- Better service: Retailers have to be competitive in terms of price and service.

- Employment: Increased employment opportunities in terms of widening the range of players involved- economists, financial personnel, bankers, market traders, brokers, in addition to the classical power engineers, arising from greater business and industry investment.

5. AFTER-EFFECTS OF DEREGULATION

The deregulation process is still under transition and a clear picture of the structure has not emerged in many countries as of now[9]. This section attempts to examine very roughly how deregulation has affected the power system in general, with particular reference to the Swedish deregulation process. It is to be noted that the Swedish deregulation is just about five years old now and hence drawing conclusive views from these discussions may be misleading. However, the following distinct features after deregulation of the Swedish market in 1996, are noteworthy.

Figure 7 shows the market prices for the Swedish electricity market since it joined the NordPool in 1996. It is evident that the market prices were very high during 1996. This can be attributed to two factors- (a) it was a very dry year in terms of rainfall (b) the market competition was meager (initially there were only a few participants). Over the years, with increasing competition, the prices have gradually reduced and have more or less stabilized.

Also, the price variations during a year reveal that the summer prices in Sweden are drastically low compared to the winter prices. This is due to the reduced electricity demand in summer. During winter, the system load is sometimes twice as much as that during summer since large heating loads come up. The prices in the market follow the load pattern too.

We observe that since 1997 and till December 2000, the monthly average price has remained stabilized within an upper limit of SEK 150/MWh and a lower limit of SEK 75/MWh, except for a few months. However, it should be noted that the prices shown here are monthly average prices. The actual hour-to-hour market price could have been vastly different and beyond these limits.

[9] As of January 2001

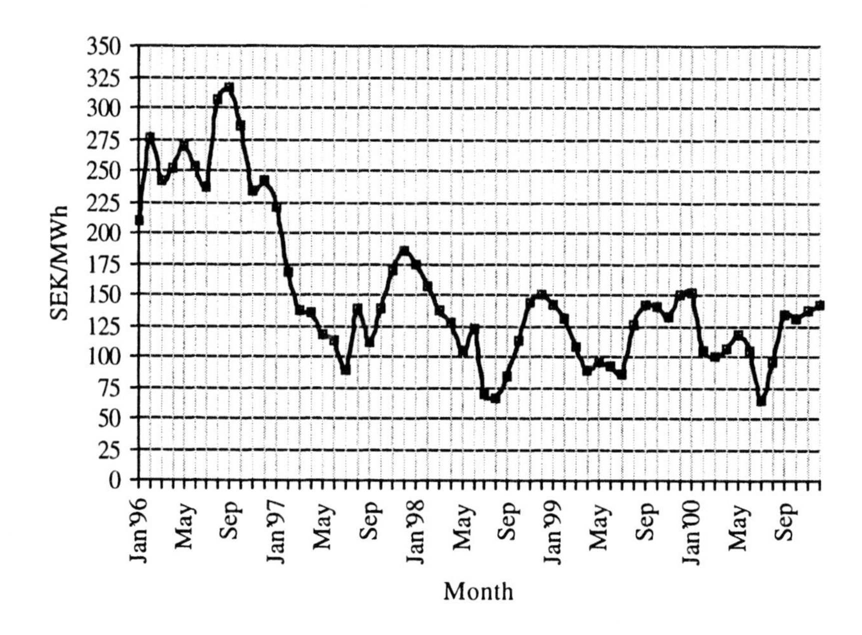

Figure 7. Spot market price in NordPool for the Swedish electricity market since its deregulation in 1996. (1 US$ ≅ 10 SEK as in January 2001) Source: NordPool ASA, http://www.nordpool.no

Another significant trend in Sweden has been the closing down of several generating units belonging to private power producers in recent years. As seen from *Figure 8*, this phenomenon started from 1997 and during 1998, installed capacity to the order of 2000 MW was shut down. Most of these units were gas based or condensing power units and were primarily being used for peak hour generation, and had high operating costs. The 1999 data however, also includes a 600 MW nuclear unit that was shut down based on a national policy of phasing out all nuclear units by 2010. This was not a fall out of the deregulation process.

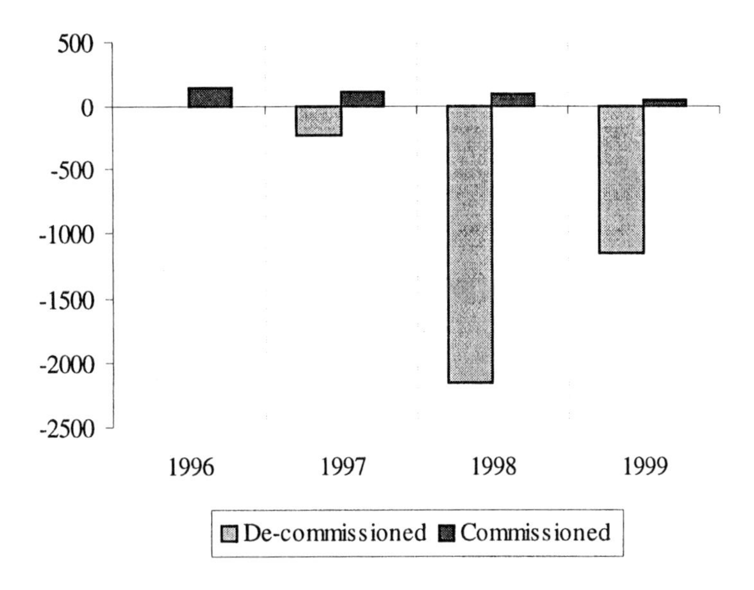

Figure 8. Sweden: Costly units can't sustain Source: Swedish National Energy Administration (http://www.stem.se)

Figure 9 shows a forecast over the next five years, regarding the total generating capacity and estimated peak load in the Swedish system. There is very little, if any, new generating capacity planned during this period, while some more generating capacity is scheduled to be shut down. The net result is a negative trend in installed capacity. Coupled with this, there is a demand growth, which should be met. All these make the scenario quite grim, since the capacity margins by the year 2005 will be very small.

Note that in deregulated electricity markets generation planning and investment decisions are prerogatives of the private investors. With spot market prices often showing a downward decline due to increased competition, a good internal rate of return from a new project is often difficult. This has an adverse impact on generation capacity addition in the system and leads to operating the system with very low security margins.

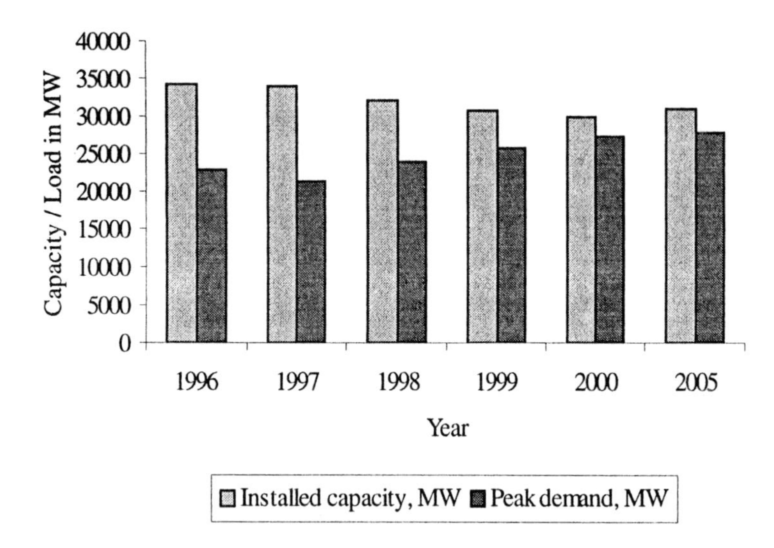

Figure 9. Sweden: The margins are reducing (years 2000 and 2005 are prognosis data)
Source: Svenska Kraftnät (http://www.svk.se) and Swedish National Energy Administration (http://www.stem.se)

Table 1 shows some of the grid specific performance indicators of the system since 1993, *i.e.,* from three years prior to deregulation. It is seen that the number of outages and the consequent total undelivered energy increased significantly since 1996. This has been a noticeable aspect in the Swedish system, however whether the disturbances can be attributed to the power sector deregulation process as such, is yet to be found out or examined.

Table 1. An overview of the performance of the Swedish national grid during the recent few years. The shaded region denotes the period after deregulation

Year	Number of disturbances during the year	Number of disturbances which led to power outage	Total undelivered energy during the year, MWh	Average interruption time[10], sec/year
1993	175	4	12	3
1994	184	1	1	0
1995	196	5	6	2
1996	162	7	133	37
1997	291	9	279	78
1998	207	8	84	24
1999	228	10	96	27

Source: Various Reports of Svenska Kraftnät and Swedish National Energy Administration

[10] Average failure time is calculated for the system as a whole by considering the total energy not supplied during the year.

These have been some of the observations pertaining to Sweden since deregulation took place in 1996. Though there have not been any major system disturbances, the operating margins have reduced considerably and the future situation seems to be even more difficult. In addition the spot market prices have not been favorable for investors in new generating companies to survive.

6. CONCLUDING REMARKS

The electric power industry is undergoing vast changes in many parts of the world. Restructuring of the industry involves a transition from natural monopolies with centralized planning to markets that are subject to competition. Work is in progress in the European Union for creating an internal market with increased competition and free pricing which could contribute towards increased consumer welfare. The United States has introduced various guidelines for the free access of the transmission wires to the participating market players. Restructuring has been taking place there, across the country, at varying pace and in varying models. While many of the countries in South America have found substantial returns from restructuring their power sector, the same cannot be said uniformly for all other regions. Many of the Asian nations are currently restructuring their power sector, and understandably there has been significant turmoil in the process.

However, it can be safely said that deregulation of the electric power industry has come to stay and the way the power industry is operated will never be the same again. This raises a whole new set of problems and issues which will be both a challenge and an experience for the power engineers of the coming years.

REFERENCES

1. (US) Energy Information Administration, "The Restructuring of the electric power industry: A capsule of issues and events", 2000.
2. (US) Federal Energy Regulatory Commission, "Promoting wholesale competition through open access non-discriminatory transmission services by public utilities: recovery of stranded costs by public utilities and transmitting utilities", Order No. 888, Issued April 24, 1996.
3. (US) Federal Energy Regulatory Commission, "Open access same-time information system (formerly, real-time information networks) and standards of conduct", Order No. 889, Issued April 24, 1996.
4. (US) Federal Energy Regulatory Commission, "Regional transmission organizations", Order No.2000, December 1999.

5. H. M. Merrill, "Regional transmission organizations: FERC Order 2000", IEEE Power Engineering Review, July 2000, pp.3-5.
6. The European Commission, *Opening up to choice: The single electricity market*, 1999.
7. R. D. Tabors, "Lessons from the UK and Norway", IEEE Spectrum, Aug.'96, pp.45-49.
8. Svenska Kraftnät, "The Swedish electricity market reform and its implications for Svenska Kraftnät", Second Edition, 1997.
9. National Electricity Market Management Company (Australia), "An introduction to Australia's National Electricity Market", 1998.
10. W. Mielczarski, G. Michalik and M. Widjaja, "Bidding strategies in electricity markets", Proceedings of Power Industry Computer Applications Conference, 1999.
11. T. Alvey, D. Goodwin, X. Ma, D. Streiffert and D. Sun, "A security constrained bid clearing system for the New Zealand wholesale electricity market", IEEE Transactions on Power Systems, May '98,pp.340-346.
12. J. Parikh, K. Bhattacharya, S. Reddy and K. Parikh, *Chapter 5- Energy System: Need For New Momentum*, India Development Report: 1997, Oxford University Press, 1997.
13. A. K. David, "Risk modeling in energy contracts between host utilities and BOT plant investors", IEEE Transactions on Energy Conversion, June 1996, pp.359-366.
14. H. Rudnick, "Pioneering electricity reform in South America", IEEE Spectrum, Aug.'96, pp.39-44.
15. H. Rudnick, "The electricity market restructuring in South America: Successes and failures on market design", Paper presented at the Plenary Session of Harvard Electricity Policy Group, Jan 1998, California.
16. Kirit Parikh, "The Enron story and its lessons", Journal of International Trade and Economic Development, July 1997, Vol.6, pp.209-230.
17. A. J. Wood and B. F. Wollenberg, *Power Generation, Operation and Control*, John Wiley and Sons, 1996.
18. F. C. Schweppe, M. C. Caramanis, R. D. Tabors and R. E. Bohn, *Spot Pricing of Electricity*, Kluwer Academic Publishers, 1988.
19. M. Einhorn and R. Siddiqi, *Electricity Transmission Pricing and Technology*, Kluwer Academic Publishers, 1996.
20. Marija Ilic, Francisco Galiana and Lester Fink (Edited), *Power Systems Restructuring: Engineering and Economics*, Kluwer Academic Publishers, 1998.
21. G. Zaccour, *Deregulation of Electric Utilities*, Kluwer Academic Publishers, 1998.
22. M. Shahidehpour and M. Marwali, *Maintenance Scheduling in Restructured Power Systems*, Kluwer Academic Publishers, 2000

FURTHER READING

In this section we briefly point out to some of the important text-books which have dealt with power sector deregulation and associated topics from a broad framework.

Wood and Wollenberg's text [17] provides a very good understanding of power system operation and control problem for classical system operation in vertically integrated environment. The basic text which introduced the concept of nodal pricing and the theory of spot pricing, and subsequently laid the foundation for electricity sector restructuring was by Schweppe *et al.* [18]. In [19] the economic and technical aspects associated with power transmission in deregulated electricity market has been addressed. The comprehensive text by Ilic, Galiana and Fink [20] covers a very wide range of topical issues on deregulation, including those on operations, control and planning in the new environment. The regulatory issues, such as compensation for stranded costs, efficiency gains, institutional design and pricing, have been dealt with, by Zaccour in [21]. Long and medium term issues in operations planning, including maintenance scheduling, in a restructured environment has been treated in [22].

LIST OF RELATED WEBSITES

http://www.eia.doe.gov	US Energy Information Administration
http://www.ferc.fed.us	US Federal Energy Regulatory Commission
http://www.nerc.com	North American Electricity Reliability Council
http://europa.eu.int/comm/eurostat	Energy related statistics of EU
http://www.m-co.co.nz	New Zealand market operator
http://www.nyiso.com	New York ISO
http://www.pjm.com	PJM Interconnection, US
http://www.nordpool.no	Nordic market operator
http://www.svk.se	Swedish ISO
http://www.fingrid.fi	Finnish ISO
http://www.statnett.no	Norwegian ISO
http://www.eltra.dk	Danish ISO
http://www.stem.se	Swedish National Energy Administration
http://www.ofgem.gov.uk	UK regulator
http://www.nemmco.com.au	Australian market operator and ISO
http://www.transpower.co.nz	New Zealand ISO
http://cercind.org/	Central Regulatory Authority of India
http://www.caiso.com/	California ISO
http://www.nationalgrid.com/	UK pool operator and ISO

Chapter 2

Power System Economic Operation Overview

1. INTRODUCTION

Power system operation in many electricity supply systems worldwide, has been experiencing dramatic changes due to the ongoing restructuring of the industry. The visible changes have been many, shifting of responsibilities, changes in the areas of influence, shift in the operating objectives and strategies, distribution of work, amongst others.

This chapter looks at the basic aspects of economic operation of a power system from a classical perspective where power generation, transmission and distribution are all owned and operated by a single entity. The objective of the system operator, in such a scenario, is to satisfy the system load in the best possible way, that is, in the most reliable, secure and economic manner.

In this environment, the activities of the system operator can be divided over three distinct time periods:

a) Pre-dispatch- planning activities

The pre-dispatch stage may comprise a period of about a week ahead of actual operation to a day ahead. A short-term forecast of the hourly aggregate system load is usually available to the operator, based on which the schedule for unit operation is drawn up for the plan period. This can also include scheduling of available hydro resources in the system, depending upon their reservoir levels and other factors. Also, plans for power exchange with other regions can be drawn up and included while formulating the unit operation schedules.

b) Dispatch- short-term scheduling

The dispatch stage begins about 30 minutes ahead of the actual operation and in this stage the operator carries out short-term planning activities such as system load flow and economic scheduling. The operator has a fair idea of the load expected on the system during this stage, including power exchanges scheduled, and accordingly, decides upon the necessary reactive support switching and load curtailment, if any. He is also responsible for maintaining adequate reserves in the system at suitable locations, which can be called upon at a short notice, in case of emergencies.

c) Instantaneous dispatch- activities in real-time

This stage could range from 5 minutes ahead of real time to actual operation. In this stage, the operator implements the plans chalked out at the pre-dispatch and dispatch stages. His activities include continuous improvement of pre-dispatch and dispatch decisions based on real-time data, initiating secondary frequency control actions if required and updating the generation schedule based on participation factors of the units and information on load.

A number of algorithmic analysis programs and simulation software are available to the operators to aid their decision-making process and make power system operations more economic and reliable. Depending on the relative accuracy and computational burden, these programs are used in the dispatch or pre-dispatch stages.

In the subsequent sections, the basic planning and dispatch activities of the system operator are discussed.

2. ECONOMIC LOAD DISPATCH (ELD)

2.1 The Economic Load Dispatch Problem

The ELD activity is executed in the dispatch stage and it primarily involves allocating the total load between the available generating units in such a way that the total cost of operation is kept at a minimum. An ELD is generally executed every 5 minutes, and hence it is very important that the solution algorithm used is efficient enough. On the other hand, the ELD model should also represent the system in as much detail as possible. An ELD formulation in its most elementary form is discussed below.

Let us consider a system of NT thermal generating units ($i \in 1, 2, \ldots,$ NT), each unit feeding power, P_i, to a transmission system, which serves a total load of PD. The network loss in the transmission system will be a function of line impedance and currents. The currents can be represented as function of the independent variables P. Thus, the network losses can be represented as a function of power generation, *i.e.*, $P_{Loss} = f(P)$.

The real power output of a generator depends on the fuel input and hence the heat input. The operating data from a generating unit is used to develop the unit's heat input-output characteristic. Subsequently, the cost input-output characteristic can be obtained using the heat input-output characteristic and fuel cost rates. In [1], two polynomial curve-fitting methods have been discussed for evaluating the unit cost characteristic. *Figure 1* shows a typical input-output characteristic of a thermal generating unit.

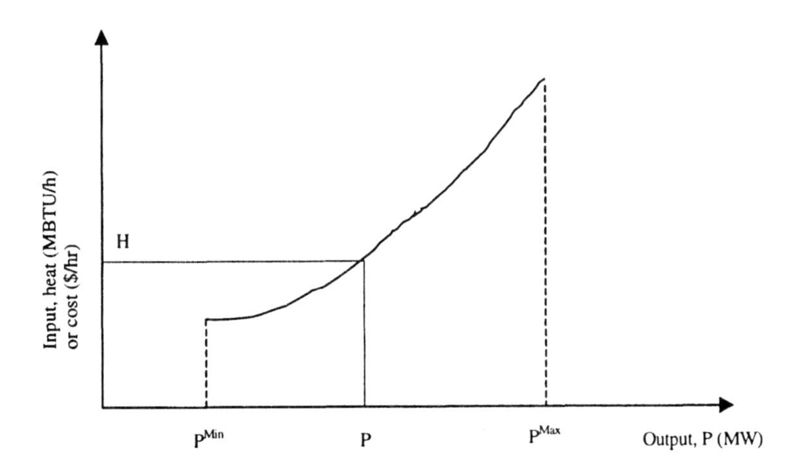

Figure 1. A typical thermal generating unit input-output characteristic

Assuming therefore, that the units' cost characteristics are available to the operator as a function of P_i, say $C_i(P_i)$, the total generating cost for the whole system is the sum of the generating cost of each generating unit. The objective of ELD is to minimize the total cost while satisfying all technical constraints.

The essential constraint on operation is that the total generation from all units must meet the total system demand and also provide for that amount of additional energy that is lost in the network due to transmission losses. Mathematically this can be written as,

$$\varphi = \sum_{i=1}^{NT} P_i - PD - P_{Loss} = 0 \tag{1}$$

Additionally, power generation from a generating unit is normally limited by an upper and lower limit respectively (2). The upper limit P^{Max} is usually decided by the de-rated capacity of the unit, while the lower limit P^{Min}, is decided from the unit's flame stability considerations.

$$P_i^{Min} \le P_i \le P_i^{Max} \qquad \forall \quad i = 1, \ldots, NT \tag{2}$$

The total system cost, which is to be minimized by the operator to obtain the optimal generation schedule, can be written as,

$$J = \sum_{i=1}^{NT} C_i(P_i) \tag{3}$$

Equations (1)-(3) constitute the ELD problem in its basic form and resemble a typical constrained optimization problem with an objective function (3), an equality constraint (1) and 2NT inequality constraints (2). This model can be solved using advanced calculus methods involving the Lagrange function as discussed in the next section.

2.2 Conditions for the Optimum

In unconstrained minimization of a function *f(x)*, the point of inflexion (the location of extremum values) can be determined from the condition, *df(x)/dx = 0*. Similarly in the case of minimization with constraints, an augmented objective function, obtained by a linear combination of the objective function and associated constraints, is formulated to obtain an optimal solution that minimizes J and also satisfies the constraints. The augmented objective function, known as *Lagrange function* or *Lagrangian,* can be formulated for the ELD optimization problem given by (1), (2) and (3) as follows:

$$\begin{aligned} F(\underline{P}, \lambda, \underline{\mu}, \underline{\gamma}) = & \sum_{i=1}^{NT} C_i(P_i) - \lambda\left[\sum_{i=1}^{NT} P_i - PD - P_{Loss}\right] - \mu_1\left[-P_1 + P_1^{Min}\right] \\ & - \mu_2\left[-P_2 + P_2^{Min}\right] - \ldots - \mu_{NT}\left[-P_{NT} + P_{NT}^{Min}\right] \\ & - \gamma_1\left[P_1 - P_1^{Max}\right] - \gamma_2\left[P_2 - P_2^{Max}\right] \\ & - \ldots - \gamma_{NT}\left[P_{NT} - P_{NT}^{Max}\right] \end{aligned} \tag{4}$$

The undetermined multipliers λ, μ_i and γ_i are known as *Lagrange multipliers*.

For the Lagrangian in (4) to have extreme values, the necessary conditions are obtained when the first derivative of F with respect to each independent variable is set to zero, *i.e.*,

$$\frac{\partial F}{\partial P_i} = 0; \quad \frac{\partial F}{\partial \lambda} = 0; \quad \frac{\partial F}{\partial \mu_i} = 0; \quad \frac{\partial F}{\partial \gamma_i} = 0 \quad \forall \ i = 1, \dots, NT \tag{5}$$

The conditions in (5) are commonly known as *Kuhn-Tucker*'s conditions of optimality. In this case there are (3NT+1) variables- the NT power outputs P, the NT multipliers each of both, μ and γ and the scalar multiplier λ. Accordingly (5) provides (3NT+1) equations in (3NT+1) variables.

The NT equations, which result when the partial derivative of F is taken with respect to P_i, are commonly known as *co-ordination equations* and are given as follows.

$$\frac{\partial F}{\partial P_i} = \frac{dC_i}{dP_i} - \lambda\left(1 - \frac{\partial P_{Loss}}{\partial P_i}\right) + \mu_i - \gamma_i = 0$$

$$\forall i = 1, 2, \dots, NT \tag{6}$$

$$\Rightarrow \quad \frac{dC_i}{dP_i} + \lambda \frac{\partial P_{Loss}}{\partial P_i} + \mu_i - \gamma_i = \lambda$$

The partial derivative of F with respect to λ in (5) merely yields the demand-supply balance of (1). The partial derivative of F with respect to μ_i and γ_i, results in 2NT equations, which are practically the inequality constraints of (2).

2.2.1 What is the Significance of the Lagrange Multipliers?

A common question and one of the significant parameters in the development of power system economics, electricity pricing theory, and the work in spot pricing theory, has been the question- what is the significance of λ, μ, and γ? We shall attempt to explain it here.

The multiplier λ appears in the Lagrange function F, associated with the demand-supply balance. It denotes the sensitivity of the objective function (cost, in this case) to a change in system demand. Simply stated, λ denotes the change in the system cost for a 1 MW change in the system demand. λ is commonly known as the system incremental cost.

The multipliers μ and γ are associated with the generator limits and these are of significance only when a generation limit is a binding constraint. In such a case, these multipliers denote the sensitivity of the objective (cost) function to the unit limits. If the generation limiting constraints are non-binding, the values of these multipliers are zero.

Due to this characteristic of μ and γ multipliers, the ELD problem is often solved without considering generation limits, so as to keep the solution easy to handle. If the generation limits are violated, then the limits and the associated μ and γ multipliers are introduced.

An important conclusion that can be reached here is that, when the generating limits are not binding, then, $\mu_i = 0$ and $\gamma_i = 0$, for all i=1,...,NT. The Lagrangian of (5) therefore simplifies to,

$$F(\underline{P}, \lambda, \underline{\mu}, \underline{\gamma}) = \sum_{i=1}^{NT} C_i(P_i) - \lambda \left[\sum_{i=1}^{NT} P_i - PD - P_{Loss} \right] \tag{7}$$

and applying the Kuhn-Tucker's condition of optimality on (7) we get,

$$\frac{\partial F}{\partial P_i} = 0 \Rightarrow \frac{dC_i(P_i)}{dP_i} = \lambda \tag{8}$$

From (8), we can conclude that when generator's limits are not binding, the optimal solution to the ELD problem is obtained when all units operate at the same value of λ, *i.e.*, *equal incremental cost operation yields the optimal dispatch.*

2.3 A Review of Recent Developments in ELD

Researchers have proposed ELD models up to various degrees of complexities and many techniques have been proposed to solve the problem more efficiently. In [2], a detailed survey has been reported for the important works in this area during the period 1977-88. For the period prior to this, comprehensive surveys carried out by Happ [3] and IEEE Working Group reports [4 and 5] are very informative.

In [6 and 7] the generation scheduling problem has been considered over a time range and the optimal dispatch is evaluated for time intervals within this range, considering load variations as well as generating unit ramping constraints. This class of problem is termed as the *dynamic economic dispatch (DED)* problem. The cost objective function is taken as a sum of the total cost over the time range. The unit ramping limits are usually of the form,

$$RDN_i \le P_i^{k+1} - P_i^k \le RUP_i \tag{9}$$

RDN and RUP are the unit ramp rates, and k denotes the time interval. This constraint makes the DED problems dynamic in nature, which makes the solution much more complex than the usual ELD.

In [8], the ELD problem is addressed for units having some prohibitive operating zones, thereby making the problem a non-convex optimization problem. In [9], the ELD problem has been addressed in the context of system reliability by including spinning reserve requirements while in [10] the problem includes reserve allocation constraint. In [11], the ELD problem has been extended to include the objective of reducing emissions from the system. The new objective function is a linear combination of the generation cost component and the cost of the emission reduction component.

2.4 Example

Consider a system with two generating units serving a total system load of 700 MW. The generator cost characteristics and operating limits are given as follows:

Table 1. Generator Characteristics

	Unit-1	Unit-2
P^{Min}	100 MW	50 MW
P^{Max}	500 MW	250 MW
Cost characteristics C_i	$a_{oi}P_i^2 + b_{oi}P_i + c_{oi}$	
a_o	1.0 \$/MW^2-h	3.4 \$/MW^2-h
b_o	8.5 \$/MWh	25.5 \$/MWh
c_o	5 \$/h	9 \$/h

Find the optimum generation level in each unit if the total system cost is to be a minimum.

Solution

The total system cost is the sum of the individual generating costs of the two units, *i.e.*, $C_1(P_1) + C_2(P_2)$; which is to be minimized subject to the constraint that the total generation meets the system demand. Let us assume that there are no transmission losses and also ignore the generating unit limits. The ELD optimization problem can then be written as follows:

$$\text{Minimize, } \sum_{i=1}^{2} C_i(P_i)$$

Subject to the constraint, $P_1 + P_2 = PD$. Using the system data, we can formulate the Lagrangian by augmenting the cost function with the demand-supply balance and an associated Lagrange multiplier λ.

$$F = (1.0P_1^2 + 8.5P_1 + 5) + (3.4P_2^2 + 25.5P_2 + 9) - \lambda(P_1 + P_2 - 700)$$

Applying the Kuhn-Tucker's conditions for optimality,

$$\frac{\partial F}{\partial P_1} = 0; \qquad \frac{\partial F}{\partial P_2} = 0; \qquad \frac{\partial F}{\partial \lambda} = 0$$

We have the following simultaneous equations in P_1, P_2 and λ:

$$2.0P_1 - \lambda = -8.5; \quad 6.8P_2 - \lambda = -25.5; \quad P_1 + P_2 = 700$$

Solution of the above equations yields the optimal solution to the ELD problem:

P_1^* = 542.841 MW
P_2^* = 157.159 MW
λ = 1094.182 \$/MWh
Cost = \$387,288.489 ≅ \$0.387 million

This is the globally optimal solution in the sense that the system cost is the least achievable, since we have not considered any other constraint. It is evident that in achieving the least-cost the optimal dispatch, the solution has violated the generating unit upper limit of unit-1. Therefore, the solution obtained above is not feasible and we need to re-work the problem by incorporating the upper limit of generation at unit-1 as an inequality constraint, *i.e.*, $P_1 \le 500$ MW. The modified Lagrangian will be of the form,

$$F = (1.0P_1^2 + 8.5P_1 + 5) + (3.4P_2^2 + 25.5P_2 + 9)$$
$$- \lambda(P_1 + P_2 - 700) - \gamma_1(P_1 - 500)$$

Applying Kuhn-Tucker's condition again, with the additional condition $\partial F/\partial \gamma_1 = 0$, and solving for the unknowns, we obtain the revised optimal solution:

P_1^* = 500 MW
P_2^* = 200 MW
λ = 1385.5 \$/MWh
γ_2 = -377.0 \$/MWh
Cost = \$395,364.0 ≅ 0.395 million = an increase of \$8076.

This increase is due to the shift of 42.841 MW of generation from unit-1 to unit-2 due to the imposition of the upper limit on unit-1. Also note that λ has increased significantly on imposition of the additional constraint.

As we mentioned earlier, the non-zero value of γ_2 (γ_2 = -377.0 \$/MWh) indicates that the upper limit of generation on unit-2 is a binding constraint and if the limit is relaxed by 1 MW the system cost will reduce by \$377.

3. OPTIMAL POWER FLOW AS A BASIC TOOL

During the last decade, the literature on Optimal Power Flow (OPF) has seen a dramatic rise, with the focus on two aspects- first, the solution methodologies, and second, the application areas. There has been a continuous search for more efficient methods for the solution of the OPF problem, more so, due to the inherent attractiveness of OPF and its promise of growth in further new application areas.

The review papers on OPF solution methodologies [12, 13] provide a good overview and discuss the relative merits of various mathematical programming techniques that have been used for solving OPF. Another comprehensive survey of OPF literature is found in [14] where the publications are classified according to the optimization techniques. As mentioned in [14], the area of OPF has, over the years, developed considerably with the emergence of sophisticated methods and network modeling techniques. Commercially available OPF codes are now able to satisfy the full ac load flow equations and can also include security and contingency constraints.

OPF was defined in the early 1960s as an extension of the conventional ELD problem to determine the optimal settings for control variables while satisfying various constraints [15, 16, 17]. In the ELD problem we discussed in Section-2, the demand was considered to be an aggregate parameter for the entire system. The inherent assumption was that the power flow follows such a simple equality constraint. In practice however, power flow follows the physical laws of electricity and network configuration, as established by the *Kirchoff's Laws*, and are commonly known as the *load flow equations*.

When these load flow equations are introduced in ELD as a system of demand-supply balance constraints, the optimum solution yields a set of decision variables satisfying the physical laws of flow of electricity while achieving a desired objective (of cost minimization, loss minimization etc.).

Such a formulation is called the *Optimal Power Flow (OPF)* and this is a static, constrained, nonlinear, optimization problem. Due to the presence of the detailed network configuration instead of the *lumped* formulation of ELD, the demand is now disaggregated and is available at all buses individually. In addition, due to the fact that the load flow equations include a reactive power balance at each node, the OPF has the additional advantage of considering reactive power as a decision variable. This single feature has led to the opening up of a large area of application, *the reactive power planning problem*.

Another notable feature of OPF is its applicability over a wide time horizon. From the system operator's perspective, an OPF needs to be executed every 30 minutes or so, to determine the optimal dispatch and control actions to be taken. OPF simulations are also necessary for medium-term planning studies, in the order of months ahead, such as reactive power planning, production scheduling, maintenance scheduling, etc. OPF also has applications in long-term studies, *viz.* generation and transmission expansion planning, which are carried out years in advance to formulate the decisions on investments in the generation and transmission system.

3.1 The Basic OPF Model

If the OPF is solved so as to minimize the total generation cost, the solution that is obtained, is a more accurate estimate than the ELD solution. The OPF objective function can however, also seek other objectives depending on the nature of the problem being addressed. For example, minimizing transmission loss is the usual objective for reactive power planning problems, or minimizing the generation shift and control actions, is used in some contingency studies. An OPF model can incorporate various control variables and system constraints as per the problem requirement. Among the control variables, an OPF set-up can include one or more of the following:

- Real and reactive power generation
- Switched capacitor settings
- Load MW and MVAr (load shedding)
- LTC transformer tap settings, etc.

We discuss below a general OPF model, which minimizes the total generation costs, subject to a set of commonly used system constraints.

3.1.1 Objective Function

A common objective function used in OPF studies is the minimization of generation costs. There may be some variations to that, for example, a component of cost denoting the operation costs associated with reactive power switching, or costs involved in load curtailment, or cost of energy not served can also be included. The objective function based on generation operating cost can be expressed as,

$$J = \sum_{i=1}^{NG} C_i(P_i) \tag{10}$$

NG is the set of all generating units including the generator on the slack bus.

3.1.2 Network Equations

The network equations are obtained from the basic *Kirchoff's Laws* governing the loop flow and nodal power balances as follows:

$$P_i - PD_i = \sum_j |V_i||V_j|Y_{i,j}\cos(\theta_{i,j} + \delta_j - \delta_i) \qquad \forall\ i = 1,\ldots,N;\ i \notin \text{slack} \qquad (11)$$

$$Q_i - QD_i = -\sum_j |V_i||V_j|Y_{i,j}\sin(\theta_{i,j} + \delta_j - \delta_i) \qquad \forall\ i = 1,\ldots,NL \qquad (12)$$

V is the bus voltage, δ is the angle associated with V, $Y_{i,j}$ is the element of bus admittance matrix, θ is the angle associated with $Y_{i,j}$, P and Q are real and reactive power generation respectively, PD and QD are real and reactive power demand respectively and NL is the number of P-Q buses.

3.1.3 Generation Limits

$$P_i^{Min} \le P_i \le P_i^{Max} \qquad \forall\ i \in NG \qquad (13)$$

$$Q_i^{Min} \le Q_i \le Q_i^{Max} \qquad \forall\ i \in NG \qquad (14)$$

P^{Min} and P^{Max} are the upper and lower limits on real power generation and Q^{Min} and Q^{Max} are the upper and lower limits on reactive generation from a unit.

3.1.4 Bus Voltage Limits

This constraint ensures that the voltages at different buses in the system are maintained at specified levels. The generator bus (or PV bus) voltages are maintained at a fixed level. Voltage level at a load bus is maintained within a specified upper limit V^{Max} and a lower limit V^{Min}, determined by the operator.

$$|V_i| = \text{constant}, \quad \forall\ i = 1, \ldots, NG$$
$$V_i^{Min} \le |V_i| \le V_i^{Max}, \quad \forall\ i = 1, \ldots, NL \tag{15}$$

3.1.5 Limits on Reactive Power Support

This constraint may be required in case the system operator has to include decisions on optimal reactive switching at load buses. Consequently, the cost objective function should be augmented with a term representing the reactive costs so as to penalize excess reactive support selection.

$$QC_i^{Min} \le QC_i \le QC_i^{Max} \qquad \forall\ i \in NL \tag{16}$$

QC^{Min} and QC^{Max} are the limits on bus reactive power support.

3.1.6 Limits on Power Flow

Transmission lines are limited by their power carrying capability, which is determined by the thermal capacity of the line or the surge impedance loading. Imposing this constraint along with (13) and (15) in the OPF ensures that the system operates in a secure manner.

$$P_{i,j} \le P_{i,j}^{Max} \qquad \forall\ \ Y_{i,j} \ne 0 \tag{17}$$

$P_{i,j}$ is the power flow over a line *i-j*. and $P^{Max}{}_{i,j}$ is the maximum limit on power flow over the line.

Following the formulation of Dommel and Tinney [16], the set of unknown variables in (11) and (12) can be identified as follows:

a. Voltage magnitude at P-Q buses
b. Voltage angle at all buses except slack bus
c. Net reactive power injection at P-V buses, *i.e.*, $(Q_i - QD_i)\ \forall\ i \in NPV$

Similarly, the specified variables in the above OPF formulation are given as follows,

a. Voltage magnitude and angle at slack bus
b. Voltage magnitude at P-V buses
c. Net real power injection at all P-V, P-Q buses, *i.e.*, $(P_i - PD_i)\ \forall\ i \notin$ slack
d. Net reactive power injection at P-Q buses, *i.e.*, $(Q_i - QD_i)\ \forall\ i \in NL$

The set of specified variables can be partitioned into those, which are controllable by the operator and those, which are fixed. For example, real power generation and voltage magnitude at PV buses can be considered controllable variables. On the other hand, voltage magnitude and angle at slack bus and net real power injection at P-Q buses would be fixed variables.

3.2 Example

Let us consider the same system of generating units described in Section 2.4 with additional information provided on the transmission network configuration. Assume that the total system load is 550 MW with associated reactive power demand of 177.45 MVAr. The load is now considered distributed over the system.

An OPF will be solved for this system to determine the optimal generation schedule while meeting the power flow constraints, system bus voltage constraints and generation limits on real and reactive power. The result will be compared with an ELD solution based on the method discussed earlier. Subsequently, we shall examine the change from the optimal generation schedule when the operator seeks to minimize system losses instead of system cost.

Figure 2 shows the system network configuration, *Table 2* provides the system data relating to generation and load while *Table 3* provides the system network data. For the sake of convenience, the generator cost characteristics used in Section-2.4 are again provided in *Table 2*. Note that bus-1 and bus-2 being generator buses (*i.e.*, PV buses), the voltages at these buses are specified exactly in *Table 2* while at other buses, the allowable upper and lower limits of voltage are specified. Also note that reactive power limits are provided at the generator buses and at buses 4 and 6, where there is a provision for capacitor support. The OPF solution also determines the optimal reactive power support from these sources.

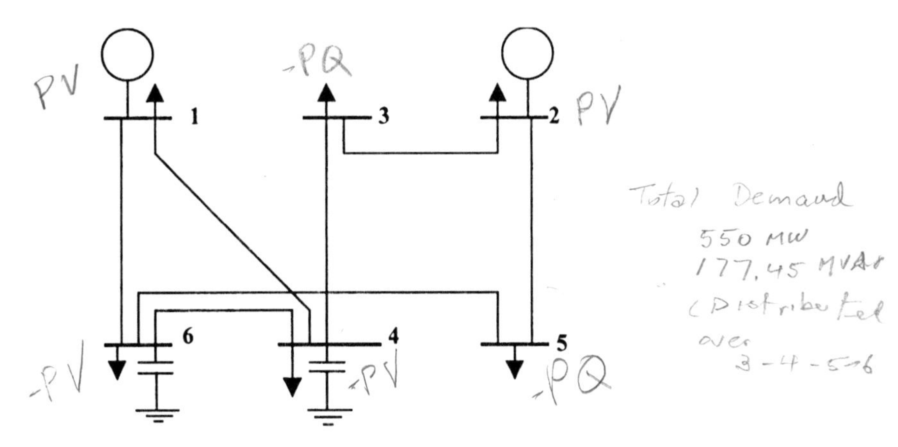

Figure 2. System configuration for the OPF simulation

Table 2. System data relating to generation and load

Bus	Generation Capacity, MW	Generator Cost Characteristic, $/hr	Load (MW + jMVAr)	Voltage, p.u.	Reactive support, MVAr
1	$100 \le P_1 \le 400$	$P_1^2+8.5P_1+5$	48.75 + j13.00	1.05	$-20 \le Q_1 \le 300$
2	$50 \le P_2 \le 200$	$3.4P_2^2+25.5P_2+9$	61.75 + j19.50	1.06	$-20 \le Q_2 \le 150$
3	NIL	-	52.00 + j26.00	$0.9 \le V_3 \le 1.1$	NIL
4	NIL	-	74.75 + j20.80	$0.9 \le V_4 \le 1.1$	$0 \le Q_4 \le 100$
5	NIL	-	84.50 + j22.75	$0.9 \le V_5 \le 1.1$	NIL
6	NIL	-	35.75 + j16.25	$0.9 \le V_6 \le 1.1$	$0 \le Q_6 \le 100$
Total			**550 + j177.45**		

Table 3. System network data

From Bus to Bus	Resistance, R (in p.u.)	Reactance, X (in p.u.)	Line Charging, $y_{i,j}/2$ (in p.u.)
1 - 4	0.0662	0.1804	0.003
1 - 6	0.0945	0.2987	0.005
2 - 3	0.0210	0.1097	0.004
2 - 5	0.0824	0.2732	0.004
3 - 4	0.1070	0.3185	0.005
4 - 6	0.0639	0.1792	0.001
5 - 6	0.0340	0.0980	0.004

Solution

(a) Cost minimization

Solving for the OPF model[1] with the cost minimization objective function, we obtain the optimal generation schedule, reactive power dispatch decisions and marginal cost of real and reactive power supply at a bus as given in *Table 4*.

Table 4. OPF solution with cost minimization objective function

Bus	Optimum Generation, MW	Optimum reactive support, MVAr	MCp, $/MWh	MCq, $/MVArh	Total system cost, $	Total system loss, MW
1	426.04	12.274	860.58	0.0	294848.378	51.955
2	175.915	113.793	1221.725	0.0		
3	-	-	1244.837	20.981		
4	-	100.00	1148.348	19.775		
5	-	-	1297.163	53.981		
6	-	100.00	1200.707	31.690		

MC_P = marginal cost of real power: Lagrange multipliers associated with (11)

MC_Q = marginal cost of reactive power: Lagrange multipliers associated with (12)

The OPF solution in *Table 4* can be compared with an ELD solution obtained using the method discussed in Section-2 (*Table 5*). Since

[1] The OPF model is solved using the MINOS-5.0 non-linear programming solver in Generalized Algebraic Modeling Systems (GAMS) programming environment [18].

transmission losses have been neglected in the ELD formulation, the total system cost is less than that in the OPF solution and cost equivalent to about 52 MW of transmission losses are not accounted for by the ELD. However, it should be noted that loss formulations could be included within the ELD model, thereby making the system representation more accurate.

Table 5. ELD solution with cost minimization objective function

Bus	Optimum Generation, MW	Incremental cost, λ $/MWh	Total system cost, $
1	426.932	862.364	240547.58
2	123.068		

Examining the two different generation schedules (*Table 4* and *Table 5*) it can be seen that the costlier generator, *i.e.*, unit-2 effectively supplies the power loss in the transmission system.

An important outcome from the OPF solution is the availability of bus-wise marginal costs for both real and reactive power. For example, the system marginal cost λ obtained from the ELD solution is 862.364$/MWh which signifies the increase or decrease in system cost if system demand increases or decreases by 1 MW respectively.

On the other hand, from an OPF solution we obtain the incremental cost at each bus, both for real and reactive power, denoted by MC_P and MC_Q respectively. These indicate the change in system cost for a 1 MW or 1 MVAr increase or decrease in real or reactive power load respectively, at that bus.

(b) Loss minimization

The transmission losses can be expressed in terms of bus voltages and associated angles as given below:

$$\text{Loss} = \frac{1}{2}\sum_i \sum_j G_{i,j}\left(V_i^2 + V_j^2 - 2 \cdot V_i \cdot V_j \cdot \cos(\delta_j - \delta_i)\right)$$

In the above, $G_{i,j}$ is the conductance of line *i-j*. *Table 6* shows the OPF solution obtained using the above loss minimization objective function. The generation schedule including the reactive power schedule changes from the earlier least-cost solution. The system cost increases while the system loss, as expected, reduces.

It is to be noted that the Lagrange multipliers corresponding to (11) and (12) now denote the change in system loss in MW, for 1 MW or 1 MVAr change in real or reactive power demand respectively, at a bus. Thus MC_P and MC_Q is replaced with a different set of parameters, namely, λ_P and λ_Q. Interestingly, these are zero for generator at bus-1 indicating that any 1 MW or 1 MVAr change in demand at bus-1, can be negotiated by the unit, without affecting the system loss (*i.e.* without increasing power flows). However, when the generation from a unit reach its limits (bus-2), the value of marginal loss parameters no longer remain at zero.

Table 6. Optimum solution with loss minimization objective function

Bus	Optimum generation, MW	Optimum reactive support, MVAr	λ_P, MW/MW	λ_Q MW/MVAr	Total system cost, $	Total system loss, MW
1	331.722	2.911	0	0	331748.042	31.722
2	250.000	92.275	0.157	0		
3	-	-	0.198	0.018		
4	-	74.717	0.209	0		
5	-	-	0.305	0.033		
6	-	99.988	0.249	0		

λ_P = marginal loss due to real power supply at a bus
λ_Q = marginal loss due to reactive power supply at a bus

3.3 Characteristic features of OPF

The main features of an OPF is its ability to include the detailed network configuration and bus-wise demand balance for both active and reactive power, within the optimization framework which, thereby, enables an exact representation of *incremental losses*. The OPF can include many operating constraints and model other issues also. Specifically- limits on reactive power generation in addition to real power generation, power flow limits (in either MW or MVA) on the transmission lines and limits on the bus voltages ensure that the system is operated in a secure manner.

3.3.1 OPF Applications

As stated earlier, one of the main features of the OPF is that it is a flexible analytical tool, which allows the use of different objective functions to solve different problems. The objective functions commonly used for operations and planning studies are as follows:

- Minimize cost of operation
- Minimize the deviation or minimize control shift
- Loss minimization
- Minimize the cost of load curtailment
- Minimize the cost of installation of new capacitors and reactors and/or cost of MVAr supplied
- Minimize the total emissions, and so on.

The above is not an exhaustive list and there has been work where combinations of the above objective functions have been used. Some of the important applications of OPF are discussed below.

3.3.1.1 Security Constrained Economic Dispatch

In the standard ELD problem, the total system cost is minimized by varying the power generation from units. In security constrained economic dispatch (SCED), an OPF is used to determine the binding network constraints and generation constraints and these are then used in the economic dispatch to evaluate the optimal generator dispatch [19]. Normally an SCED is run every 15 minutes in control centers.

3.3.1.2 Preventive and Corrective Rescheduling

Preventive rescheduling control actions are initiated when the contingency analysis detects a contingency problem in the system. In such circumstances, the OPF can be used to evaluate the optimal set of control actions required both during pre- and post-contingency conditions [20]. Minimization of cost is the usual objective function.

Corrective rescheduling comprises those actions taken by the operator to remove security violations such as a line overload present in the system, as soon as possible [21]. Minimum operating cost, minimum number of controls or minimum shift from the optimum operation can be used as the objective function. The optimization function should be run often and should use only the control necessary to eliminate the violations.

3.3.1.3 Reactive Power Planning and Voltage Control

System planning studies are carried out to determine the best location and size of capacitors or other reactive power devices to be placed in the system so that the system variables, such as bus voltages, are within specified limits. In order to achieve an optimal choice of investment decisions, OPF has been successfully used in many instances [22, 23, 24, 25, 26]. It is the prerogative of the planner to include different constraints, including those on post-contingent states to find the desired location of these devices in the system. The objective functions used for determining the optimal choice also varies. While an objective function to minimize transmission losses will result in a reactive plan with a good voltage profile on the system, the cost considerations in such optimal plans are also of importance. In many cases, the objective is to minimize the cost, which includes a term representing the annualized cost of investment.

Voltage control is generally carried out by equipment switching at the local level. However, in certain cases, system-wide OPF can also be used for determining voltage control decisions [27]. Another application is demonstrated in [28] wherein a linear programming based OPF is used to reschedule active power controls to correct voltage magnitude constraint violations. Though under normal conditions, equipment switching would be the preferred method, this method may be applicable as a corrective step when the system is close to voltage collapse.

3.3.1.4 Power Wheeling and Wheeling Loss Calculation

Power wheeling is an arrangement between two parties for buying and selling of energy, using the transmission network owned by a third party (wheeling utility). Several issues come up for analysis when power wheeling is involved in a system and OPF has been shown to be very effective for such analysis. Among the many issues that an OPF can examine, include:

- Is there enough transmission capacity in the wheeling utility system?
- Is there a need to provide additional reactive support at certain buses to enhance the transfer capability of the system? If so, at which nodes and how much should be charged for providing that support? Who should be charged for this service?
- Is there a need to re-dispatch the generating units in the remaining units in the system to accommodate the wheeling transaction? If so, what are the changes, which units need rescheduling and what is the cost of these changes?
- Is the system secure enough to carry out the transaction?
- How much of the transmission capacity of the wheeling utility is being utilized? This provides the basis for determining the wheeling rate.
- How much loss is incurred in wheeling and which party bears this burden?
- How to determine the short-term marginal cost of wheeling?

An OPF model can be used to address these issues and also determine what actions need to be taken so that overloading on critical transmission lines is avoided and reliability levels are maintained [29, 30, 31].

3.3.1.5 Pricing of Real and Reactive Power

The principle of spot pricing of electricity was introduced in [32], which subsequently lead to major changes and restructuring of the power industry. With the help of OPF models, it is possible to develop pricing mechanisms for both real and reactive power delivered at a bus in the system.

In mathematical terms, the 'delivery' of real or reactive power at a bus implies the algebraic sum of the real or reactive power flow on the lines incident on the bus including real or reactive generation injections or load drawn from the bus. The real power price based on the marginal cost at a bus can be given by,

$$\rho_{P_i} = MCp_i - \lambda_i^{Min} + \lambda_i^{Max} \qquad (18)$$

λ^{Min} and λ^{Max} are duals associated with the constraints on lower and upper generation limit respectively, while MCp is the marginal cost of real power delivery at a bus [33]. Similarly, the reactive power price at a bus can be determined by,

$$\rho_{Q_i} = MCq_i - \mu_i^{Min} + \mu_i^{Max} \tag{19}$$

μ^{Min} and μ^{Max} are the duals associated with the constraints on lower and upper reactive generation limit respectively while MCq is the marginal cost of reactive power delivery at a bus [33, 26]. It is to be noted that the above formulation for real and reactive power prices can be extended to include the duals of other operating constraints as well.

4. UNIT COMMITMENT (UC)

The UC problem is a somewhat longer term scheduling problem usually covering a time range from 24 hours (1 day) to 168 hours (1 week) ahead, and is handled by the operator in the pre-dispatch stage. In this problem, the operator needs to take decisions on how to commit or de-commit (*i.e.* keep running or shut down) its available units over the week, or over the next day. The input to the operator is the demand forecast for the next week or next day, as the case may be, aggregated for the whole system.

In the same way, that the ELD problem is formulated, the operator seeks to minimize its system costs over the planning horizon in a UC problem, while meeting the forecast demand to decide upon the unit up / down status for every hour.

This planning activity is essential due to the fact that the system load varies over a day or even over a weekly period and hence it is not economical to keep all the units on-line for the entire duration. A proper schedule for starting up, or shutting down the units can save costs significantly.

UC problems are much more complex to solve, compared to the ELD problem discussed earlier, due to the presence of binary decision variables on unit status (on /off). Depending upon the need of the system and computational facilities, the utilities choose to use UC models that suit their requirements.

In the following sub-sections, we first discuss the basic structure of an UC model and subsequently some of those additional issues addressed within the UC set up by various researchers.

4.1 UC: The Basic Model

4.1.1 Objective Function

As mentioned earlier, the operator's objective function, while solving the UC problem still remains the same, *i.e.*, to minimize the system costs. However, due to the longer time-scale of the problem, the total system cost will be affected by the start-up and shut down decisions of generating units, effected within the planning period. These are explained below.

- Fuel Cost: There have been two different approaches to represent fuel costs in UC models. The first and the most common approach has been to use a cost characteristic derived from the heat-rate characteristics and represented by a polynomial function, which is usually quadratic, and can be written as follows,

$$C_i = a_i P_i^2 + b_i P_i + c_i \qquad (20)$$

  The other approach has been to represent the generator cost as a constant, which is derived from the generator's *average full load cost.*

- Start Up Cost: This component appears in the UC objective function to take into account the cost incurred during a generator start-up operation. This is often modeled as a function of the time for which the unit was off-line.

$$ST_i = \alpha_i + \beta_i \left(1 - e^{-T_i^{OFF}/\tau_i}\right) \qquad (21)$$

  α is a fixed cost associated with the unit start-up, β is the cost involved in a cold start-up, T^{OFF} is the time for which the unit has been off and τ is a time-constant representing the cooling speed of the unit.

  Another approach has been to use a constant cost representation, which is included in the objective function when the unit is on start-up.

- Shut Down Cost: Usually this component of cost is not considered in UC models since it is not very significant compared to other costs. However, a constant cost representation has been used in [34], and is included when the unit undergoes a shut down.

The composite objective function for the UC problem can be constructed using the above as follows:

$$J = \sum_{k=i}^{T} \left[C_{i,k}(P_{i,k}) \cdot W_{i,k} + ST_{i,k} \cdot UST_{i,k} + SD_{i,k} \cdot USD_{i,k}) \right] \tag{22}$$

UST, USD and W are integer decision variables denoting the status of the unit at hour *k*. W denotes the unit status (1 = running, 0 = off); UST denotes the unit start-up state (1 = start-up, 0 = no start-up) and USD denotes the unit shut down state (1 = shut down, 0 = no shut down).

4.1.2 Constraints in Unit Commitment

4.1.2.1 Demand-Supply Balance and Spinning Reserves

This constraint ensures that the operator has scheduled enough capacity for a particular hour so that the demand at that hour is met. This can also include any pre-decided import or export contracts with other utilities, and a certain amount of reserve capacity. A typical demand-supply balance constraint is given as follows.

$$\sum_{i=1}^{NG} P_{i,k} \cdot W_{i,k} + (I_k - E_k) \geq PD_k + RESV_k \tag{23}$$

The term RESV denotes the *spinning reserve* in the system, which is a reserve available to the operator from among its spinning units, *i.e.* from the generators already running. Therefore, this reserve is available almost instantaneously to the operator in case of need. The operator has the very important responsibility of maintaining adequate spinning reserves in the system, not only on a total-MW basis, but he also needs to take care of the location aspect of this reserve, taking into account transmission capacities available in the system.

The operator generally uses his experience or certain rules for determining this reserve to be maintained in his system. As mentioned in [35], the RESV component could typically comprise a base component, a fraction of the load requirement and a fraction of the high operating limit of the largest on-line unit.

4.1.2.2 Minimum Up and Down Time Constraints on Thermal Units

These constraints ensure the minimum number of hours a unit must be on, before it can be shut down (minimum up-time) or the minimum number of hours a unit must be off-line before it can be brought on-line again (minimum down-time). These constraints are usually applicable to large thermal units.

$$\sum_{n=1}^{MUT} USD_{th,k-n+1} \leq 1; \qquad \forall \ k \geq MUT$$

$$\sum_{m=1}^{MDT} UST_{th,k-m+1} \leq 1; \qquad \forall \ k \geq MDT \tag{24}$$

MUT is the minimum time in hours the unit should be running before shut down; MDT is the minimum time in hours the unit should remain shut down before start-up.

4.1.2.3 Generation Limit

This constraint describes the allowable range of generation available for scheduling, as defined by the maximum and minimum limits of the unit.

$$P_i^{Min} \leq P_{i,k} \cdot W_{i,k} \leq P_i^{Max} \tag{25}$$

4.1.2.4 Must-Run Units

Some units such as large coal-based units or nuclear units cannot be start up or shut down on a day-to-day basis following the daily load variations because they involve very high start-up costs and other technical constraints. Such units need to be assigned a *must-run* status.

$$W_{i,k} = 1; \qquad \forall \quad i \in MR \tag{26}$$

4.1.2.5 Crew Constraints

These constraints pertain to the number of units that can be started at the same time in a particular plant.

4.2 UC: Additional Issues

The UC model described in the earlier sub-section, or one of its variations is generally used by the dispatcher to obtain the commitment schedules. Many additional issues have been addressed within the UC framework, upon which we shall dwell briefly next.

4.2.1 Multi-Area UC

UC has been adapted in the context of multi-area systems where individual utilities are interconnected by tie lines and the dispatch is carried out jointly [36, 37]. The additional constraint introduced in this class of problems is the inter-area transmission constraint, which can impose severe restrictions on the optimal solution. In [36] the inter-area transmission lines are modeled using a linear flow network model while in [37] a dc power flow representation is used. It is seen from these works that the joint scheduling of multi-area systems can bring about significant reduction in system costs. Such systems are seen to be vulnerable to transmission capacity availability, which is a critical parameter in determining the level of savings achievable from joint dispatch.

4.2.2 Fuel Constrained UC

Another important issue addressed within the UC framework, is fuel supply planning, which when properly coordinated with generation scheduling can reduce system costs significantly.

In [38], the fuel co-ordination problem has been considered over a time horizon of one-month, for three types of units (a) oil-fired (b) fuel-constrained LNG/LPG fired and (c) pumped storage hydro units. The fuel consumption is modeled as a function of power generated and unit start-up, and constrained by a fixed availability over the planning horizon. In [34,39 and 40] different kinds of fuel contracts, constraints on gas consumption and gas delivery systems have been incorporated in the daily generation scheduling program.

4.2.3 Ramp Rate Constraint on Thermal Units

Ramp rate constraints were introduced earlier in Section-2.3 in the context of the dynamic economic dispatch problem. This constraint limits the inter-hour generation change in a unit, and is particularly applicable to coal-based thermal units. While researchers have used several models of the ramp constraint, a typical formulation is shown below.

$$\begin{aligned} P_{i,k} &\leq RUP_i \cdot P_{i,k-1} \\ P_{i,k} &\geq RDN_i \cdot P_{i,k-1} \end{aligned} \tag{27}$$

RUP and RDN are the ramp-up and ramp-down constants of a unit. This constraint links the generation variable of the previous hour to that of the present hour, and hence introduces a dynamic characteristic in the UC models [34].

A viable way of dealing with ramping constraints in a dynamic programming algorithm has been developed in [41]. In [42] the ramp rate constraints are applied exogenous to the UC model, *i.e.,* first a UC solution is obtained without the ramp limits, thereby reducing the computation. Thereafter, a backward dynamic procedure is used to update the UC schedules taking into account the inter-hour load changes and the system's capability to respond to it. A price based UC is developed in [43] wherein the impact of ramp-rate constraint is examined via the hourly marginal ramp-rate values, *i.e.* the Lagrange multipliers associated with ramp-rate constraint. In [44], ramp constraints have totally been done with, and instead, the objective function is augmented with an additional term, the ramping cost, which is related to depreciation in shaft life. While [43] has used Lagrange multipliers for each unit's ramp-rate constraint, in [45], a *system ramp multiplier* has been introduced that is based on the inter hour demand fluctuation, adjusted to obtain the optimal dispatch.

4.2.4 Transmission Constraints

Accounting for transmission losses is a subject where conventional UC models make simplifying assumptions. Similarly, most UC models neglect the power transmission limits of lines, limits on bus voltages and limits on reactive generation. It might so happen that a UC solution obtained without considering these constraints couldn't be implemented in the actual system, because it failed to satisfy the load flow requirements.

Therefore, inclusion of transmission constraints in UC programs helps to represent the transmission losses more accurately and also ensures that the actual dispatch does not deviate much from the UC solution obtained in the pre-dispatch stage.

DC load flow representations have been used in [34, 46 and 47] to represent the transmission line capacity limits though the ideal way would be to include an ac load flow model within the UC. That would however, make the computations extremely complex. In [48] a generalized UC problem has been developed that considers the power flow constraints, line flow limits as well as voltage limits. Recently, OPF with transmission security and voltage constraints have been included within the UC model formulation and is solved by decomposing the model into a UC master problem and a transmission and reactive power sub-problem [35].

4.2.5 Environmental Issues

Maintenance of environmental standards and meeting targets for reduction of emission are important considerations to those utilities with a dominant share of fossil fuel. Such utilities are required to carry out their generation scheduling activity keeping environmental constraints into account. This issue has been addressed in [35 and 47] wherein the generator's emission characteristic is represented by a polynomial function of the power generation. The total emission from the system is constrained by a pre-defined *emission cap*.

The comprehensive survey of the UC literature in [49] provides a summary of the important work, which appeared till 1991, and also classifies them on the basis of the techniques used to solve the UC problem. A clear trend has been established in the choice of the Lagrangian relaxation technique as the most preferred and efficient technique.

5. FORMATION OF POWER POOLS

So far, we discussed the basic ideas of least-cost system operation and short-term generation scheduling. The objective of the planner or system operator was to achieve the system least cost operating strategy while meeting the demand and other constraints.

In order to achieve further reduction in system cost many utilities in US during the seventies and eighties used to form power pools and operate in a coordinated manner so as to minimize the pool costs. This was possible by transmitting power from a utility, which had cheaper sources of generation to another utility having costlier generation sources. The total reduction in system cost was termed as pool savings and was shared by the participating utilities.

This form of co-operation between utilities to reduce system cost while increasing their individual benefit, was further boosted by the (US) Power Utilities Regulatory Policy Act of 1978 (PURPA). This Act required the utilities to interconnect with co-generating plants or NUG[2], to purchase power at prices up to its avoided cost. With the introduction of this Act, co-generation units received an incentive for producing additional power, something that was, earlier, not lucrative enough, and the quantity of power

[2] Non-utility generation or NUG was the term used to classify co-generating units and all such units, which satisfied the PURPA *minimum efficiency* and *fuel use* criteria. Such units were called *qualifying facilities*. However, NUG could also include large independent power producers, which provided capacity sales to utilities [50].

sold back to utilities by co-generators increased significantly. For the utilities, the participation of NUG provided much more flexibility in their least-cost operation planning and generation scheduling activities. Introduction of power wheeling mechanisms, which involve transport of power over a third party's network, further aided to achieve better economic efficiencies in the system.

5.1 Power Pools

Power pools came into existence during the seventies and eighties as a reaction to increasing prices of electricity. The objective was to reduce the overall operating cost of utilities by sharing the cheaper resources. Many different forms of power pools existed and these could be identified by the following criteria-

a. What role did the pool operator play in day-to-day operation of the pool?
   The role of the pool operator varied widely across the systems. In some, the pool operator was responsible for unit commitment and dispatch in addition to setting up the inter-utility transactions while in some others the operator also managed maintenance scheduling functions and long-term expansion planning decisions.
b. How were the transactions set up?
   Many power pools had a system of joint dispatch of generation for all areas simultaneously, to determine the optimal transactions. In other cases, they formed a brokerage system and transactions were settled based on a bidding system. The power pools also differed on how the system cost savings was allocated to various participants.

Depending on these criteria, the various structures of the power pools, which came into existence, could be grouped in to three main types:

- Economic exchange of energ
- Multi-area joint dispatch
- Energy brokerage systems

5.1.1 Economic Exchange of Energy

This is a simple arrangement for exchange of energy between two utilities having significant difference in their marginal operating costs. The utility with the higher operating cost purchases power from the utility with low operating costs. This arrangement is usually on an hour-to-hour basis and is conducted by the two system operators. The benefit from such an arrangement is ensured if the savings in production cost from two co-operating utilities, is more than the cost incurred from payments towards charges for transmission of the energy.

A modeling framework has been developed in [50] to evaluate the potential for economic energy transfers in interconnected systems. In [51] an algorithm has been developed to evaluate the inter-utility power purchases where the objective is to minimize the total generation and power purchase costs.

5.1.2 Multi-area Joint Dispatch

In multi-area joint dispatch all participating utilities coordinate to dispatch their generation in a centralized manner. In certain instances, for example, the New England power pool, the pool also coordinated the system expansion and maintenance planning activities [50].

Consider that a set of NU utilities interconnected by transmission lines; cooperate to set up a power pool. Each utility *m* ($m \in NU$) has a set of generating units NG_m with their respective generating cost characteristic, say $C(P_{i,m})$, where $P_{i,m}$ is the power generated by unit *i* located in utility *m*. In the case of joint dispatch, the objective of the central dispatcher is to minimize the total system cost subject to meeting the system constraints. We define the total system cost as,

$$J = \sum_{m}\sum_{i} C(P_{i,m}) \qquad \forall\ i = 1,\ldots, NG_m \qquad \text{and } m = 1,2,\ldots\ldots, NU \tag{28}$$

The system constraints that need to be satisfied by the model include a set of demand-supply balance equations (29) and a constraint on the transmission capacity (30). The demand-supply balance for each utility *m* can be written as,

$$\sum_{i} P_{i,m} + \sum_{k} T_{k,m} = PD_m \tag{29}$$

$T_{k,m}$ is the transfer of energy from utility *k* to utility *m*. The transmission limits can be given as,

$$T_{k,m} \le T_{k,m}^{Max} \tag{30}$$

The above representation of power transmission between utilities (29) is of the simplest form and resembles the transportation model, the only requirement being that the algebraic sum of the power at a node is zero. In a more elaborate formulation, the transmission can be represented using a dc power flow model as given below [52, 53, 54, 55].

$$T_{k,m} = (\delta_k - \delta_m) \cdot \frac{1}{x_{k,m}} \tag{31}$$

δ is the angle associated with the voltage at a bus node and $x_{k,m}$ is the reactance of the line in per unit.

An ideal representation of the transmission system would be to include an ac load flow, which however, increases the computational burden significantly.

In [56 and 56], joint dispatch models have been developed which are based on the linear flow network representation. The models are solved to obtain the expected energy generation, the energy not served and operating cost savings for each area participating in joint dispatch. The savings are allocated to each area based on a *split-the savings equally* criterion which will be discussed later.

From the objective function (28) and demand-supply constraint (29), we can formulate the Lagrangian as follows:

$$F = \left[\sum_m \sum_i C(P_{i,m})\right] - \left[\sum_m \lambda_m \left(\sum_i P_{i,m} + \sum_k T_{k,m} - PD_m\right)\right] - \left[\sum_k \sum_m \gamma_{k,m}\left(T_{k,m} - T_{k,m}^{Max}\right)\right] \tag{32}$$

The Lagrange multipliers in (32), λ_m ($m = 1, \ldots, NU$) and $\gamma_{k,m}$ ($k = 1, \ldots, NU$ and $m = 1, \ldots, NU$) provide useful information about the pool operation. A high value of λ for a utility *m* denotes a large cost saving to the system when demand of utility *m* reduces by 1 MW, *i.e.* when utility *m* imports power from another utility. On the other hand, γ denotes the worth of a transmission transaction. Therefore, those transactions aiding in reducing the system cost would have a large and negative value associated with γ.

5.1.2.1 Example

For the four-utility joint dispatch configuration shown in *Figure 3*, let us assume that each utility can be represented by a gross generation capacity and a composite cost function as per the details given in *Table 7.*

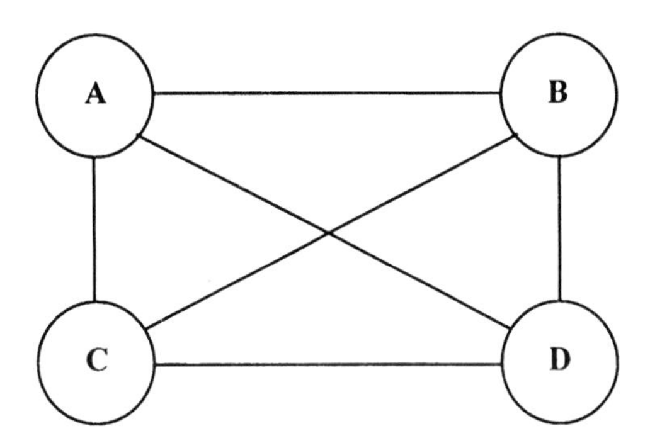

Figure 3. A four-utility joint dispatch model

Table 7. A multi-utility power interchange system

Utility	a_o, \$/MWh^2	b_o, \$/MWh	c_o, \$	P^{Max}, MW	P^{Min}, MW	PD, MW
A	1.8	10.5	0.5	150	20	120
B	2.8	24.5	0.8	250	30	200
C	3.0	15.6	0.4	230	40	180
D	1.5	20.1	0.6	125	25	75

The composite cost characteristic for each utility is assumed to be of the form:

$$C_m = a_{o_m} P_m^2 + b_{o_m} P_m + c_{o_m}$$

a_o, b_o and c_o are the appropriate cost coefficients. For the above multi-area system, find the system cost reduction when the utilities operate through a joint dispatch as compared to when they operate independently.

Solution

(a) Independent Operation:

First let us consider that the utilities operate independently and meet their own area demand from their respective sources. Following is the optimal solution:

Table 8. Optimal solution without inter-utility transfers

	P, MW	λ, \$/MWh	System cost
A	120	442.5	\$254035.3
B	200	1144.5	
C	180	1095.6	
D	75	245.1	

As seen from *Table 8*, when there is no power interchange between the utilities, they meet their respective demand through their own resources. The system cost, which is the sum of the individual utility costs, is $254035.3.

From the table we also note that, a 1 MW reduction in demand in B will reduce the system cost by $1144.5, that in C by $1095.6, and so on.

However it should be noted that, if a power transfer of 1 MW from another utility brings about the reduction in demand of utility B, the reduction in system cost would not be the same. In such a case, the cost reduction will depend on which area is providing this power to B, and is discussed in the following cases.

(b) $T_{k,m}$ = 1 MW:

Let us now consider that each of the transmission lines have a power carrying capacity of 1 MW. We assume that the central dispatcher optimizes the generation schedule so as to minimize the total system costs, while meeting the system constraints. Also note that transmission losses have been neglected here. Following is the optimal solution (*Table 9*):

Table 9. Optimal solution with a transfer capacity of 1 MW on each line

Utility	P, MW	λ, $/MWh	γ, $/MWh	System cost
A	121.0	446.1	γ_{AB} = -681.6 γ_{AC} = -643.5	$2507270.5
B	197.0	1127.0	All zero	
C	179.0	1089.6	γ_{CB} = -38.10	
D	78.0	254.1	γ_{DA} = -192.00 γ_{DB} = -873.60 γ_{DC} = -835.5	

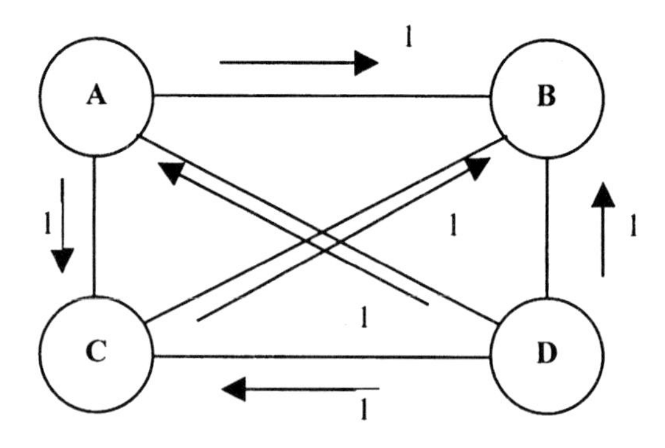

Figure 4. Power transactions in joint dispatch with 1 MW of capacity on each line

From the optimal solution and the optimal inter-utility transactions shown in *Figure 4*, we see that there are six transactions scheduled by the central dispatcher as follows: A to B, A to C, C to B, D to A, D to B and D to C; each of 1 MW.

From the generation schedules, we see that utility A increases its generation by 1 MW and utility D by 3 MW from the case (a) dispatch solution, while B reduces its generation by 3 MW and C by 1 MW. The total system cost savings from a joint dispatch and a 1 MW transfer capacity on lines is, as much as, $3307.8.

The fourth column of *Table 9* indicates the worth of a transaction. For example, $\gamma_{A,C}$ = -643.5$/MWh signifies that an additional 1MW transfer from A to D would further reduce system cost by $643.5.

It is interesting to note three different values of γ for utility D, γ_{DA}=-192.0, γ_{DB}=-873.6 and γ_{DC}=-835.5. The different values of γ shows the difference in benefits of power exports from D to different utilities. The most valuable transaction, from the total system cost viewpoint, is evidently when D exports power to B while the least valuable is when D exports power to A.

Finally, we also note that $\gamma_{B,k}$ is zero for all *k*, implying thereby that there is no benefit to the system in terms of costs savings, when B exports power.

(c) $T_{k,m}$ = 10 MW:

Now lets consider a transmission capacity of 10 MW on all inter-utility transmission lines and examine the optimum interchange schedules obtained through joint dispatch.

Table 10. Optimal solution with a transfer capacity of 10 MW on each line

Utility	P, MW	λ, $/MWh	γ, $/MWh	System Cost
A	130	478.5	γ_{AB} = -526.53 γ_{AC} = -526.53	224721.75
B	175.1	1005.03	All zero	
C	164.9	1005.03	All zero	
D	105	335.1	γ_{DB} = -669.93 γ_{DC} = -669.93	

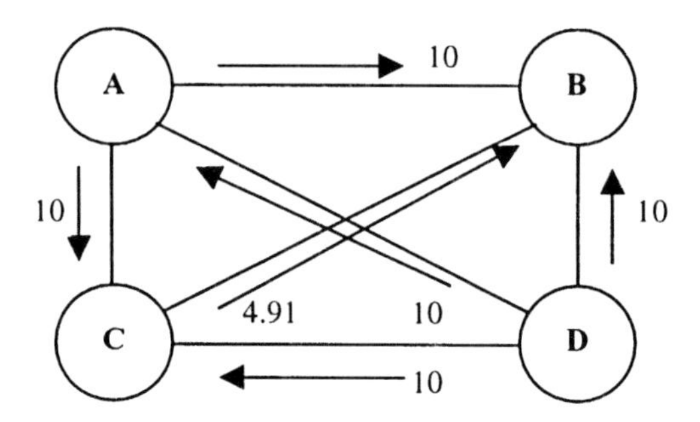

Figure 5. Power transactions in a joint dispatch with 10 MW of capacity on each line

The optimal solution in *Table 10* shows that the system cost is \$224721.75, which is about 88.5% of the base case cost reported in (a).

The utilities A and D had earlier increased their generation by 1 MW and 3 MW respectively, when the transmission limit was 1 MW. With a transmission limit of 10 MW, they proportionately increase their generation by 10 MW and 30 MW respectively. This 40 MW increased generation in A and D is compensated by reduced generation in B and C to the order of 24.9 MW and 15.1 MW respectively. *Figure 5* shows the transactions scheduled in the joint dispatch case to minimize total system costs.

(c) As Transmission Capacity Increases:

Ideally, the power pool operation will realize its maximum benefit when there are no restrictions on power transfer across utilities. However, there is a limit up to which a utility can support another utility by way of power transfers due to its own generating capacity limits. This is reflected in the plot of system cost as a function of transmission capacity, shown in *Figure 6*. As transfer capacities of inter-utility transmission lines increase, the total system cost reduces due to optimal utilization of cheaper generation. However, beyond a certain capacity of transmission, there is no further reduction in system cost. Note that the system cost has been normalized with respect to the base-cost of the system, *i.e.* with the cost of independent operation.

In this example the cost reduces to 80.5% of the base case cost if the transmission limit on each inter-utility interconnection is 20MW. The total cost reduction is by \$49605.552, which remains the same for a transmission capacity of 20 MW or more, including the case with no transmission constraints.

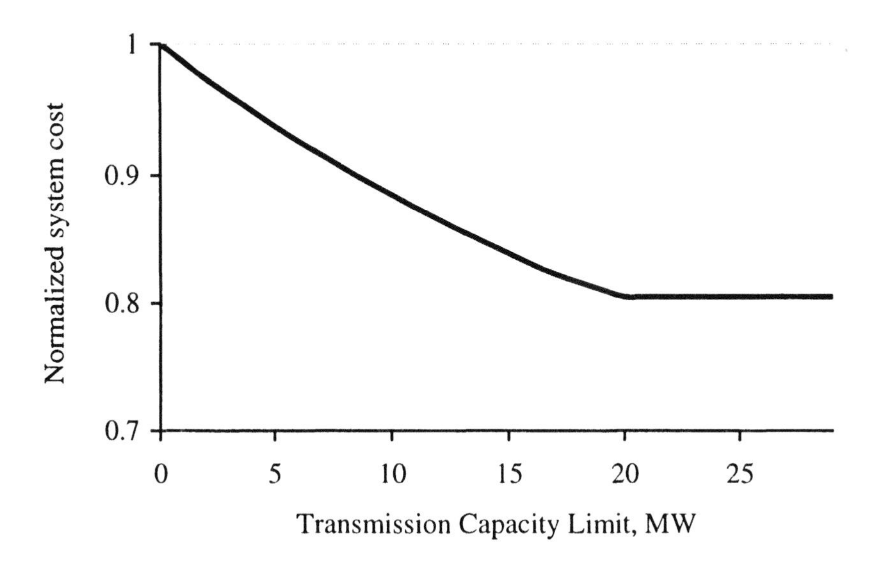

Figure 6. System cost as a function of transmission capacity

5.2 The Energy Brokerage System

In an energy brokerage system, several utilities join together to form a pool, with a central broker in place, to co-ordinate the operations on an hour-to-hour basis. The individual utilities carry out their own generation scheduling activities, unlike as in the multi-area joint dispatch case, and formulate their purchase and sell decisions. These purchase and / or sell decisions are submitted to the central broker in the form of *bids*, specifying the quantity available for purchase or sell along with the price quotes.

The sell and purchase bids are usually the incremental cost and quantity to be sold or decremental cost and quantity to be purchased, respectively. The energy broker determines the interchange schedule based on the bids, using certain criterion, for example, maximizing the social benefit.

The savings achieved from the transactions are then divided among the participating utilities. The advantage of this arrangement is that the broker can observe all purchase and sell offers simultaneously to allocate the transactions and thus achieve better economy of operation.

5.2.1 The Bids

The price-quantity bid offers that are submitted by an individual utility can be determined using ELD calculations of the type discussed in Section-2.1. First, the marginal cost is obtained from the dual of the demand-supply constraint of the ELD model. If the utility's generator cost functions are linear, the marginal cost curve is available as a piece-wise linear step function. On the other hand, if the cost functions of the generators are non-linear, say quadratic; the utility's marginal cost characteristic will comprise piece-wise linear straight lines with certain slopes. In such cases, piece-wise constant approximation for the marginal cost curve can be made. In [58], an OPF formulation has been used instead of the usual ELD, to determine the information input required in an energy brokerage system (*i.e.*, the utility's incremental cost, and amount of power available for trading).

Once the marginal cost characteristic is obtained, the utility can formulate its bid blocks, depending on its demand at an hour vis-à-vis the marginal cost. For example, if the demand at an hour is low, its marginal cost would also be low, and hence it would not be profitable to buy power at that hour. On the other hand, this situation is favorable for selling power. The utility should bid for power-sell at that hour, the quantity to sell and the price, can be determined from the marginal cost characteristic.

5.2.2 Transfer Schedule and Allocation of Savings

Once all the utilities have submitted their bids for purchase or sell to the central broker, the broker organizes the buy bids in decreasing order of bid prices and the sell bids in increasing order of their bid prices (*Figure 7*). The optimum transaction schedule is obtained by matching the highest priced buy bid with the lowest priced sell bid. This method of matching the bids has been practiced in many brokerage systems in US though it has been shown in [59] that under certain conditions when there are some contract restrictions between utilities, the method can give sub-optimal schedules.

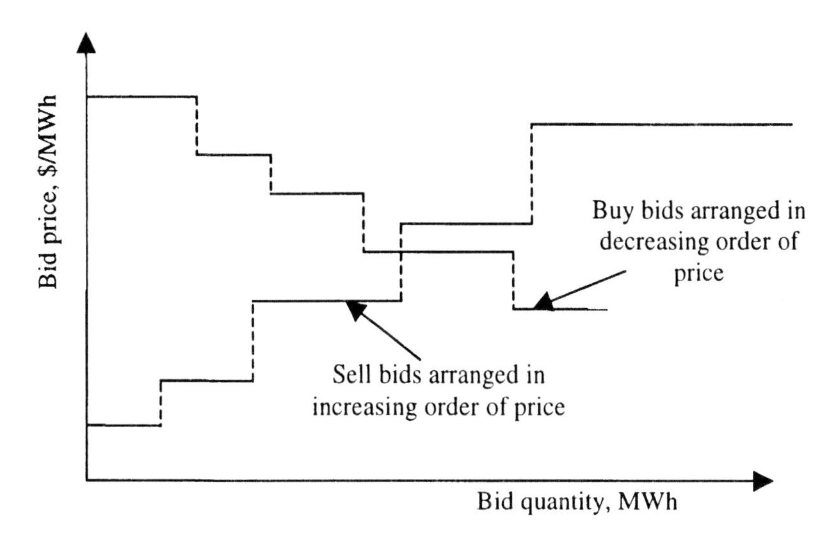

Figure 7. Power purchase and sell bids in energy brokerage systems

The optimal transaction schedules can also be obtained in an optimization framework where the objective function is modeled as the broker's profit function defined as the difference between buy bid price and sell bid price [60].

$$J = \sum_{k} \sum_{m} \left(C_{Buy_m} - C_{Sell_k} \right) \cdot T_{k,m} \tag{33}$$

In (33), C_{Buy} is the buy bid price of utility *m* and C_{Sell} is the sell bid price of utility *k* and $T_{k,m}$ is the power transfer from utility *k* to *m*. The objective function in (33) is the sum of the savings accrued from each high-low match and is denoted as the broker's profit (or system savings). This function is maximized subject to the transmission capacity constraints of the lines.

Several methods for the allocation of system savings have been proposed and applied to real brokerage systems. A good discussion is provided in [59] about the merits and demerits of some of the most common methods, namely, the 'split the savings', pool averaging, proportional participation and proportional contribution. A brief outline of the method of *split the saving equally* is given below.

5.2.3 Split the Savings Equally

When the transactions are finalized by the broker, the price $\rho_{T_{i,j}}$ is fixed for each transaction based on the buyer's decremental cost and seller's incremental cost, as given by:

$$\rho_{T_{k,m}} = \frac{1}{2} \cdot \left(C_{Buy_m} + C_{Sell_k}\right) \tag{34}$$

The buying utility's savings, S_{Buy}, is accounted for, from the reduction of generation cost net of payment made to the selling utility while the selling utility's savings, S_{Sell} comes from the additional revenue it receives from the buying utility net of its increased generation cost.

$$S_P = \left(C_{P_m} - \rho_{T_{k,m}}\right) \cdot T_{k,m} \tag{35}$$

$$S_S = \left(\rho_{T_{k,m}} - C_{S_k}\right) \cdot T_{k,m} \tag{36}$$

The total system savings from a transaction, $S=S_{Buy}+S_{Sell}$, is given by:

$$S = \frac{1}{2} \cdot \left(C_{Buy_m} - C_{Sell_k}\right) \cdot T_{k,m} \tag{37}$$

5.2.4 Example

Consider the example system used in Section-5.1.2.1. Now, instead of a central dispatcher scheduling the generation of all the utilities, we assume that there is a central broker who receives bids for sell and purchase of energy from the four utilities. The broker schedules the interchanges by matching the highest buy bid with the lowest sell bid. Find the interchange schedules and savings to the system.

<u>**Solution**</u>

The utilities simulate their ELD individually and obtain their respective marginal cost characteristic, which is the first derivative of the cost characteristic, and can be found to be as follows, for the given system:

$$\frac{dC_A}{dP_A} = 3.6P_A + 10.5 \qquad \frac{dC_B}{dP_B} = 5.6P_B + 24.5$$

$$\frac{dC_C}{dP_C} = 6.0P_C + 15.6 \qquad \frac{dC_D}{dP_D} = 3.0P_D + 20.1$$

Accordingly, the plot of the above characteristics is shown in *Figure 8*.

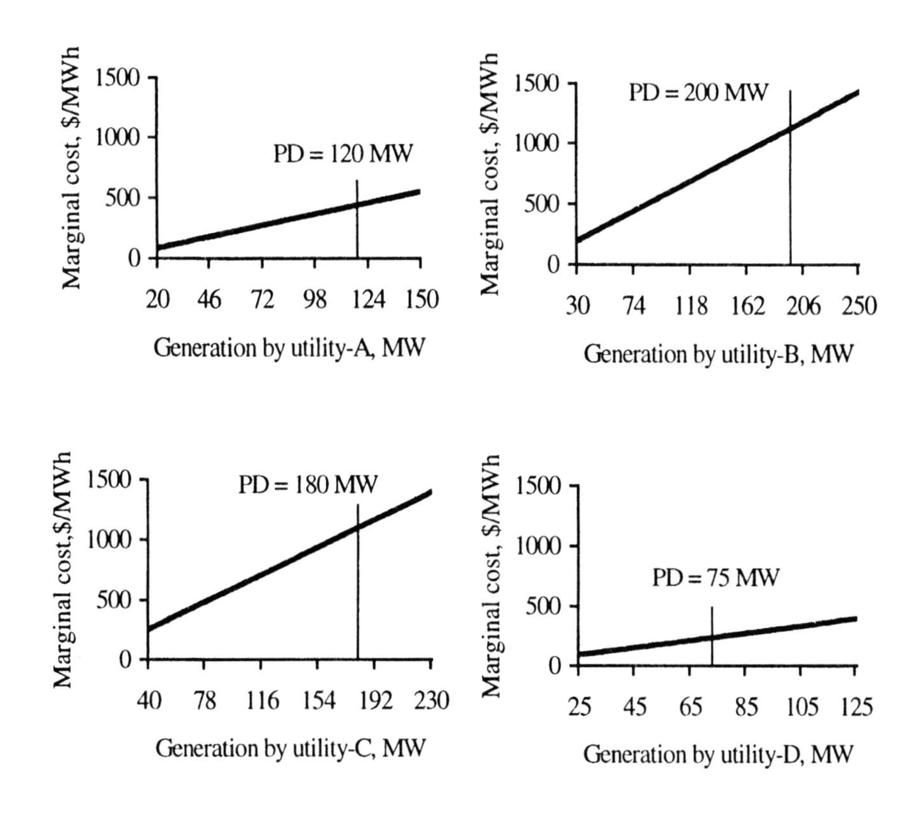

Figure 8. Marginal cost characteristics of each utility. The vertical line shows the power demand in each utility area.

We assume that the utilities place their bids without considering unit commitment decisions. Therefore, for example, if a utility so desires, it can purchase as much energy as to meet all of its demand and shut down its generators. Based on the marginal cost characteristic, step-wise bid blocks for buy and sell can be formulated for each utility. *Figure 9* shows the formulation of such bid blocks for utility-A. Since the marginal cost characteristic is a monotonically increasing linear function, a band size of 50 MW has been used to construct the bid blocks. It is to be noted that such a bid formulation is not a unique one.

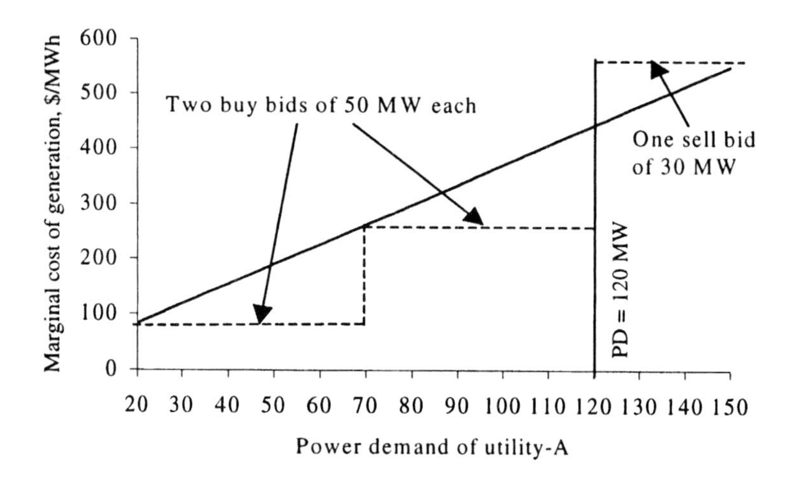

Figure 9. Marginal cost characteristic of Utility-A and formulation of its bidding blocks

In this particular instance, utility A has a demand of 120 MW, which is shown by the vertical line. The marginal cost corresponding to this load is $442.5/MWh. Understandably, utility A will be willing to buy power at prices less than $442.5/MWh and sell power at prices more than $442.5/MWh. It can sell a maximum of 30 MW, after meeting its load of 120 MW, since it has a maximum generation capacity limit of 150 MW. Similarly, if we assume that it will not shut down all its own generating units, it can buy power up to 100 MW while generating the minimum 20 MW, to meet its load. Accordingly, the buy and sell bid blocks are prepared by considering the purchase of 100 MW in two bid blocks of 50 MW each, and the sell of 30 MW, in one bid block only and placing them at discrete price intervals.

Table 11 lists all the bid blocks thus formulated for each utility in the same manner as discussed for utility-A. Utility B can sell a maximum of 50 MW due to its capacity limit and thus its sell bid is in two blocks of 25 MW each, while it can buy 170 MW, which is constructed in four blocks. Utility C has two selling bid blocks and three buying bid blocks while, utility D has two blocks each for buying and selling bids.

Table 11. Bid blocks

Utility	P^{Max}, MW	P^{Min}, MW	Demand, MW	Marginal cost of the utility to meet the demand with own generation λ $/MWh	Buy Bid Quantity MW	Buy Bid Price $/MWh	Sell Bid Quantity MW	Sell Bid Price $/MWh
A	150	20	120	442.5	50	262.5	30	550.5
					50	82.5		
B	250	30	200	1144.5	50	864.5	25	1284.5
					50	584.5	25	1424.5
					50	304.5		
					20	192.5		
C	230	40	180	1095.6	50	795.6	25	1245.6
					50	495.6	25	1395.6
					40	255.6		
D	125	25	75	245.1	25	170.1	25	320.1
					25	95.1	25	395.1

The bids so formulated are submitted to the central broker for settlement of the transactions, who then arranges the bids in order, *i.e.* buy bids are arranged in descending order of prices and sell bids in the ascending order (*Table 12*).

Table 12. Buy and sell bids in decreasing and increasing order of prices respectively

Buy Bids Utility	Buy Bids Price, $/MWh	Buy Bids Quantity, MW	Sell Bids Utility	Sell Bids Price, $/MWh	Sell Bids Quantity, MW
B	864.5	50	D	320.1	25
C	795.6	50	D	395.1	25
B	584.5	50	A	550.5	30
C	495.6	50	C	1245.6	25
B	304.5	50	B	1284.5	25
A	262.5	50	C	1395.6	25
C	255.6	40	B	1424.5	25
B	192.5	20			
D	170.1	25			
D	95.1	25			
A	82.5	50			

(a) Settlement of transactions without considering transmission limits:
The bids are matched in the order in which they have been arranged in *Table 12*. Accordingly, the optimal transaction schedule is obtained as shown in *Table 13*. The broker clears three transactions in all, two of which are between the same utilities (B and D). Hence, in physical terms there are two transactions, a 50 MW transfer from D to B and a 30 MW transfer from A to C. The system savings achieved from the brokerage is $32698. This is somewhat less compared to the system savings obtained with multi-area joint dispatch case without transmission limits ($49605.5). We should note

here that the savings achieved from the brokerage system operation depends entirely on the buy and sell bids placed by the participants- over which the central dispatcher (or broker) has no authority.

Table 13. Transaction price, quantity, and savings accrued from each transaction

Transaction	C_{Buy}, $/MWh	C_{Sell}, $/MWh	Transaction Cleared, MW $T_{i,j}$	Broker's Profit or Savings, $ $(C_{Buy} - C_{Buy})*T_{i,j}$	Transaction Price $\rho_{Ti,j}$, $/MWh
B buys from D	864.5	320.1	25	13610	592.30
B buys from D	864.5	395.1	25	11735	629.80
C buys from A	795.6	550.5	30	7353	673.05
Total				32698	

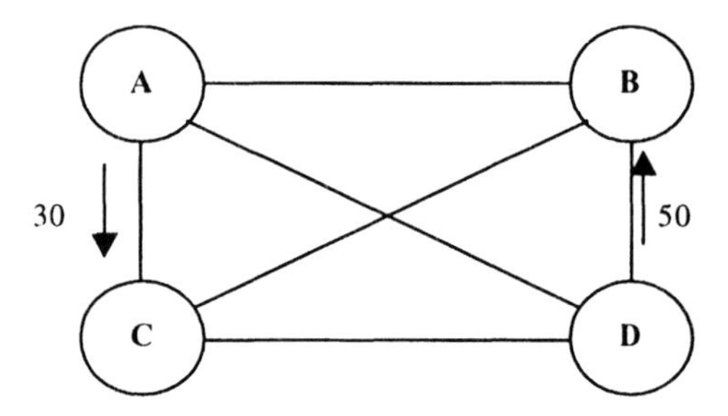

Figure 10. Optimum transaction scheduled by broker without considering transmission limits

(b) With a transmission limit of 10MW:
When the broker has to incorporate transmission limits into consideration while matching the bids, the solution is entirely different. The optimum transaction schedule and corresponding transaction prices are shown in *Table 14.*

Table 14. Transaction price, quantity, and savings accrued from each transaction

Transaction	C_{Buy}, $/MWh	C_{Sell}, $/MWh	Transaction cleared, Ti,j, MW	Broker's profit or Savings, $, $(C_{Buy} - C_{Buy})*T_{i,j}$	Transaction price, $\rho_{Ti,j}$, $/MWh
B buys from D	864.5	320.1	10	5444	592.30
B buys from A	864.5	550.5	10	3140	707.50
C buys from D	795.6	320.1	10	4755	557.85
C buys from A	795.6	550.5	10	2451	673.05
Total				15790	

Figure 11 shows the power transactions cleared by the broker. It is noted that, comparing this to the corresponding case of multi-area joint dispatch solution, there are two transactions, C to B and D to A, which do not figure in the present solution. The reason being that the lowest sell bid price of C is higher than the highest buy bid price of B, and the same applies to the case

of D to A. Therefore, there is no match possible. The saving obtained in this case is significantly less compared to the previous case, without transmission limits. However, it is to be noted that the saving is comparable to the corresponding case with multi-area joint dispatch.

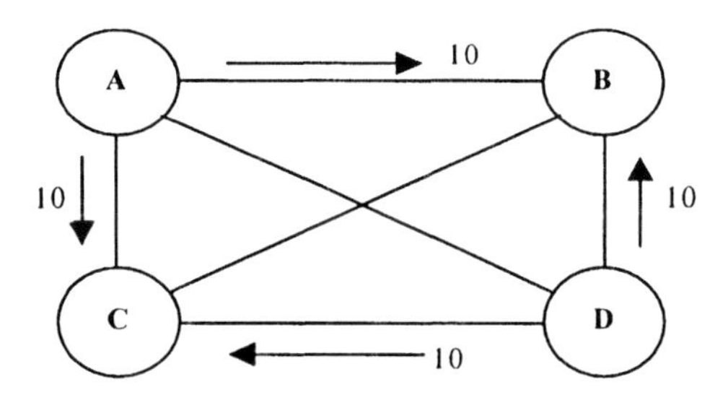

Figure 11. Optimum transaction scheduled by broker when all lines have a limit of 10 MW. (Note that this solution is similar to the joint-dispatch transactions)

6. CONCLUDING REMARKS

Power system economic operation in the vertically integrated mode has over the years involved an overall systems approach to analysis of the costs and treatment of the various issues. The system operator or planner seeks to achieve the optimal operating strategy that is beneficial to the entire system, including the generator, transmission and distribution sectors. This chapter has focused on short-term operations planning issues and covers the planning horizon from about one week ahead, when unit commitment decisions are made, to near real-time, when economic load dispatch is carried out. The benefits accrued in terms of overall cost savings, by cooperating with other utilities through economic transactions and multi-utility coordinated operations have been brought out. Such cooperation between power utilities to form power pools and energy brokerage systems, where the central broker determines the optimal trades based on bids submitted by participating utilities, can be thought of as the transitory phases from the wholly vertically integrated systems to deregulated markets.

REFERENCES

1. Z. X. Liang and J. D. Glover, "Improved cost functions for economic dispatch computations", IEEE Transactions on Power Systems, May'91, pp.821-829.
2. B. H. Chowdhury and S. Rehman, "A review of recent advances in economic dispatch", IEEE Transactions on Power Systems, Nov. '90, pp. 1248-1259.

3. H. H. Happ, "Optimal power dispatch- A comprehensive survey", IEEE Transactions on Power Apparatus and Systems, Vol.PAS-96, May / June 1977, pp.841-854.
4. IEEE Working Group, "Description and bibliography of major economy - security functions. Part-II: Bibliography (1959 - 1972)", IEEE Transactions on Power Apparatus and Systems, Vol. PAS-100, Jan.'81, pp.215-223.
5. IEEE Working Group, "Description and bibliography of major economy - security functions. Part-III: Bibliography (1973 - 1979)", IEEE Transactions on Power Apparatus and Systems, Vol. PAS-100, Jan.'81, pp.224-235.
6. C. B. Somuah and N. Khunaizi, "Application of linear programming re-dispatch technique to dynamic generation allocation", IEEE Transactions on Power Systems, Feb.'90, pp. 20-26.
7. G. Demartini, G. P. Granelli, P. Marannino, M. Montagna and M. Ricci, "Coordinated economic and advance dispatch procedures", IEEE Transactions on Power Systems, Nov. '96, pp. 1785-1791.
8. F. N. Lee and A. M. Breipohl, "Reserve constrained economic dispatch with prohibited operating zones", IEEE Transactions on Power Systems, Feb.'93, pp. 246-254.
9. N. A. Chowdhury and R. Billinton, "Risk constrained economic load dispatch in interconnected generating system", IEEE Transactions on Power Systems, Nov.'90, pp.1239-1247.
10. G. B. Sheble, "Real-time economic dispatch and reserve allocation using merit order loading and linear programming rules", IEEE Transactions on Power Systems, Nov.'89, pp.1414-1420.
11. J. Nanda, L. Hari and M. L. Kothari, "Economic emission load dispatch with line flow constraints using a classical technique", IEE Proceedings, Generation, Transmission and Distribution, Jan.'94, pp.1-10.
12. J. A. Momoh, M. E. El-Hawary and R. Adapa, "A review of selected optimal power flow literature to 1993. Part-I: Non-linear and quadratic programming approaches", IEEE Transactions on Power Systems, Feb.'99, pp. 96-104.
13. J. A. Momoh, M. E. El-Hawary and R. Adapa, "A review of selected optimal power flow literature to 1993. Part-II: Newton, linear programming and interior point methods", IEEE Transactions on Power Systems, Feb.'99, pp. 105-111.
14. M. Huneault and F. D. Galiana, "A survey of the optimal power flow literature", IEEE Transactions on Power Systems, May '91, pp. 762-770.
15. J. Carpentier, "Contribution e l'étude do Dispatching Economique", Bulletin Society Francaise Electriciens, August 1962.
16. H. W. Dommel and W. F. Tinney, "Optimal power flow solutions", IEEE Transactions on Power Apparatus and Systems, October 1968, pp.1866-1876.
17. A. J. Wood and B. F. Wollenberg, *Power Generation, Operation and Control*, 2nd Edition, John Wiley and Sons, Inc. 1996.
18. GAMS Release 2.25, *A User's Guide*, GAMS Development Corporation.
19. R. Lugtu, "Security constrained dispatch", IEEE Transactions on Power Apparatus and Systems, Jan/Feb.'79.
20. A. Monticelli, M. V. F. Pereira and S. Granville, "Security constrained optimal power flow with post-contingency corrective rescheduling", IEEE Transactions on Power Systems, February 1987.
21. B. Stott, O. Alsac and A. J. Monticelli, "Security analysis and optimization", Proceedings of the IEEE, December 1987, pp.1623-1644.
22. G. Opoku, "Optimal power system VAR planning", IEEE Transactions on Power Systems, Feb. '90, pp. 53-60.

23. Y. T. Hsiao, C. C. Liu, H. D. Chiang and Y. L. Chen, "A new approach for optimal VAR sources planning in large scale electric power systems", IEEE Transactions on Power Systems, Aug. '93, pp. 988-996.
24. N. Deeb and S. M. Shahidehpour, "Cross decomposition for multi-area optimal reactive power planning", IEEE Transactions on Power Systems, Nov. '93, pp. 1539-1544.
25. S. Granville and M. C. A. Lima, "Application of decomposition techniques to VAR planning: Methodological and computational aspects", IEEE Transactions on Power Systems, Nov. '94, pp. 1780-1787.
26. D. Chattopadhyay, K. Bhattacharya and J. Parikh, "Optimal reactive power planning and its spot pricing. An integrated approach", IEEE Transactions on Power Systems, Nov.'95, pp. 2014-2020.
27. M. Bjelogrlic, M. S. Calovic, P. Ristanovic and B. S. Babic, "Application of Newton's optimal power flow in voltage / reactive power control", IEEE Transactions on Power Systems, Nov.'90, pp.1447-1454.
28. D. S. Kirschen and H. P. Van Meeteren, "MW / voltage control in a linear programming based optimal power flow", IEEE Transactions on Power Systems, May '88, pp.481-489.
29. R. Mukerji, W. Neugebauer, R. P. Ludorf and A. Catelli, "Evaluation of wheeling and non-utility generation (NUG) options using optimal power flow", IEEE Transactions on Power Systems, Feb.'92, pp.201-207.
30. Y. Z. Li and A. K. David, "Optimal multi-area wheeling", IEEE Transactions on Power Systems, Feb.'94, pp.288-294.
31. Y. Z. Li and A. K. David, "Wheeling rates of reactive power flow under marginal cost pricing", IEEE Transactions on Power Systems, Aug.'94, pp.1263-1269.
32. M. C. Caramanis, R. E. Bohn and F. C. Schweppe, "Optimal spot pricing. Practice and Theory", IEEE Transactions on Power Systems, Sept.'82, pp.3234-3245.
33. M. L. Baughman and S. N. Siddiqi, "Real time pricing of reactive power: Theory and case study results", IEEE Transactions on Power Systems, Feb.'91, pp.23-29.
34. S. Ruzic and N. Rajakovic, "A new approach for solving extended unit commitment problem", IEEE Transactions on Power Systems, Feb. '91, pp. 269-277.
35. H. Ma and S. M. Shahidehpour, "Unit commitment with transmission security and voltage constraints", IEEE Transactions on Power Systems, May '99, pp. 757-764.
36. F. N. Lee and Q. Feng, "Multi-area unit commitment", IEEE Transactions on Power Systems, May'92, pp.591-599.
37. F. N. Lee, J. Huang and R. Adapa, "Multi-area unit commitment via sequential method and a DC power flow network model", IEEE Transactions on Power Systems, Feb. '94, pp. 279-287.
38. K. Aoki, M. Itoh, T. Satoh, K. Nara and M. Kanezashi, "Optimal long-term unit commitment in large scale systems including fuel constrained thermal and pumped storage hydro", IEEE Transactions on Power Systems, Aug.'89, pp. 1065-1073.
39. F. N. Lee, "A fuel-constrained unit commitment method", IEEE Transactions on Power Systems, Aug. '89, pp. 1208-1218.
40. S. Vemouri and L. Lemonidis, "Fuel constrained unit commitment", IEEE Transactions on Power Systems, Feb. '92, pp. 410-415.
41. W. J. Hobbs, G. Hermon, S. Warner and G. B. Sheble, "An enhanced dynamic programming approach for unit commitment", IEEE Transactions on Power Systems, Aug. '88, pp. 1201-1205.
42. C. Wang and S. M. Shahidehpour, "Effects of ramp-rate limits on unit commitment and economic dispatch", IEEE Transactions on Power Systems, Aug. '93, pp. 1341-1350.

43. F. N. Lee, L. Lemonidis and K. C. Liu, "Price based ramp-rate model for dynamic dispatch and unit commitment", IEEE Transactions on Power Systems, Aug.'94, pp.1233-1242.
44. C. Wang and S. M. Shahidehpour, "Optimal generation scheduling with ramping costs", IEEE Transactions on Power Systems, Feb. '95, pp. 60-67.
45. S. Y. Lai and R. Baldick, "Unit commitment with ramp multipliers", IEEE Transactions on Power Systems, Feb. '99, pp. 58-64.
46. J. Batut and A. Renaud, "Daily generation scheduling optimization with transmission constraints: A new class of algorithms", IEEE Transactions on Power Systems, Aug.'92, pp. 982-989.
47. S. J. Wang, S. M. Shahidehpour, D. S. Kirschen, S. Mokhtari and G. D. Irisarri, "Short-term generation scheduling with transmission and environmental constraints using an augmented Lagrangian relaxation", IEEE Transactions on Power Systems, Aug.'95, pp.1294-1301.
48. R. Baldick, "The generalized unit commitment problem", IEEE Transactions on Power Systems, Feb. '95, pp. 465-475.
49. G. B. Sheble and G. N. Fahd, "Unit commitment literature synopsis", IEEE Transactions on Power Systems, Feb. '94, pp. 128-135.
50. B. W. Bentley, "Integrating non-utility generation into the NEPOOL resource planning process", IEEE Transactions on Power Systems, Nov. '88, pp. 1754-1756.
51. N. S. Rau and C. M. Necsulescu, "A model for economy energy exchanges in interconnected power systems", IEEE Transactions on Power Systems, Aug. '89, pp. 1147-1153.
52. L. Zhang, P. B. Luh, X. Guan and G. Merchel, "Optimization-based inter-utility power purchases", IEEE Transactions on Power Systems, May '94, pp.891-897.
53. S. Ruzic and N. Rajakovic, "A new approach for solving extended unit commitment problem", IEEE Transactions on Power Systems, Feb. '91, pp. 269-277.
54. F. N. Lee, J. Huang and R. Adapa, "Multi-area unit commitment via sequential method and a DC power flow network model", IEEE Transactions on Power Systems, Feb. '94, pp. 279-287.
55. J. Batut and A. Renaud, "Daily generation scheduling optimization with transmission constraints: A new class of algorithms", IEEE Transactions on Power Systems, Aug.'92, pp. 982-989.
56. F. N. Lee, "Three-area joint dispatch production costing", IEEE Transactions on Power Systems, Feb. '88, pp. 294-300.
57. F. N. Lee, "A new multi-area production costing method", IEEE Transactions on Power Systems, Aug. '88, pp. 915-922.
58. G. Fahd and G. B. Sheble, "Optimal power flow emulation of interchange brokerage systems using linear programming", IEEE Transactions on Power Systems, May'92, pp. 497-504.
59. K. W. Doty and P. L. McEntire, "An analysis of electric power brokerage systems", IEEE Transactions on Power Apparatus and Systems, Feb.'82, pp.389-396.
60. G. Fahd, D. A. Richards and G. B. Sheble, "The implementation of an energy brokerage system using linear programming", IEEE Transactions on Power Systems, Feb.'92, pp. 90-96.

Chapter 3

Power System Operation in Competitive Environment

1. INTRODUCTION

In Chapter-1 we discussed the various factors and issues, that initiated the process of deregulation of the electric power industry all over the world. Subsequently in Chapter-2 the basic concepts of economic operation of the system were discussed from a classical perspective, *i.e.*, considering the utility to be operating in a vertically integrated environment. In this classical perspective, the system operator seeks to achieve economic efficiency for the system as a whole. The operating paradigm is based on achieving the system least-cost solution while meeting reliability and security requirements.

Given this background, we shall now examine the various issues related to the deregulated power industry and the new paradigms of system operation. First we shall discuss power sector deregulation in some more detail with emphasis on the two distinct structures of deregulation that have emerged. As we have mentioned earlier, several players have emerged in the deregulated electricity markets. Consequently, many of the activities of system operation have been taken over by different entities. For example, the genco, the transco and the system operator, each now have a role to play, independent of each other, while at the same time needing to coordinate their activities in order to maintain the system security and reliability.

2. ROLE OF THE INDEPENDENT SYSTEM OPERATOR (ISO)

The independent system operator (ISO) is the central entity to have emerged in all deregulated markets with the responsibility of ensuring system security and reliability, fair and equitable transmission tariffs, and providing for other system services. With differing market structures evolving in various countries, it has been noticed that based on the responsibilities assigned to them and their functional differences, ISOs could be placed in two categories.

The first and the more common one, is the pool structure in which the ISO is responsible for both market settlement including scheduling and dispatch, and transmission system management including transmission pricing, and security aspects. This has often been referred to, as the *pool model* and it exists in different forms in the UK, Australian, Latin American and some of the US markets.

The other structure is that of open access, one dominated by bilateral contracts, and can be found in the Nordic countries. In this system, bulk of the energy transactions are directly organized between the generator and the customer, and the ISO has no role in generation scheduling or dispatch and is only responsible for system operation. The role of the ISO is minimal and limited to the maintenance of system security and reliability functions. *Table 1* lists the basic differences between these two market structures and the role of the ISO in each.

Table 1. Comparison of the Two Different Market Structures

Open Access	Pool
♦ Bulk of the energy transactions are carried out as bilateral trades while there may also exist a day-ahead spot market ♦ The ISO is not responsible for market administration, generation scheduling or dispatch functions ♦ Market administration is carried out by a separate entity and participation in the market is not mandatory ♦ The ISO is responsible for system security and control, procuring necessary ancillary services ♦ Example: Nordic markets	♦ All energy transactions are carried out through the pool which may be organized through a day-ahead trading mechanism ♦ The ISO is responsible for the market settlements, unit commitment, and determination of the pool price ♦ Market administration is carried out by the ISO and participation of gencos is mandatory ♦ The ISO is also responsible for system security and control, procuring necessary ancillary services ♦ Example: UK market

In any market structure, be it a pool or bilateral contract dominated one, the ISO has three basic functions laid out for it, maintenance of system security and reliability, service quality assurance and promotion of economic efficiency and equity. To achieve these objectives, the ISO may be authorized to set the rules for transactions between suppliers and consumers, scheduling and dispatch of generators, loads and network services, and energy markets [1]. As noted in [2], there is not much difference between the two models since both the market structures recognize the need of an ISO to act as the regulator for system reliability and operations. In the first system, there is enough flexibility for participants for commercial transactions, while the players in the second system must adhere to careful co-ordination by the ISO. Various functions of the ISO that are essential for the system security and efficiency of power system operation have been discussed in [3].

Operational planning activities and different roles of the operator in *pre-dispatch, dispatch,* and *instantaneous dispatch* stages pertaining to vertically integrated electric utilities were discussed in Chapter-2. In deregulated power systems, there is a distinct difference between the operator's activities in the control room of an independent profit seeking genco vis-à-vis those in the control room of the system's independent operating authority, *i.e.* the ISO. In this section we discuss the operational planning activities of the ISO in both the market structures, pool and bilateral contract dominated, and in the following section we shall discuss the operations planning activities of an independent, profit seeking genco.

2.1 Structure of UK and Nordic Electricity Sector Deregulation

2.1.1 The UK Market

As mentioned in Chapter-1, the UK electricity market operates as a power pool where all gencos submit their price and quantity offers in half-hourly blocks for the next day to the ISO-cum-market operator, *i.e.*, pool operator[1]. These 48 bids for energy supply are then arranged in order of their offer prices and matched with the load forecast for each half-hour (*Figure 1*). The highest priced bid offer that matches the load forecast, becomes the *system marginal price* (SMP). The amount of generation to be scheduled is also decided from this and unit commitment decisions are conveyed to the generators [4].

The system price so obtained from matching the supply curve with the load forecast is basically an unconstrained dispatch. However, if there are outages and the system operator expects shortfalls on some account, the system price is modified with a *loss of load probability (LOLP)* based

[1] The National Grid Company of UK carries out these functions

component is known as *capacity payment*. This component provides an incentive to the gencos to maintain an adequate margin over the level of demand. The capacity payment component can be high when the margin narrows, but is zero if there is a large excess of generation available.

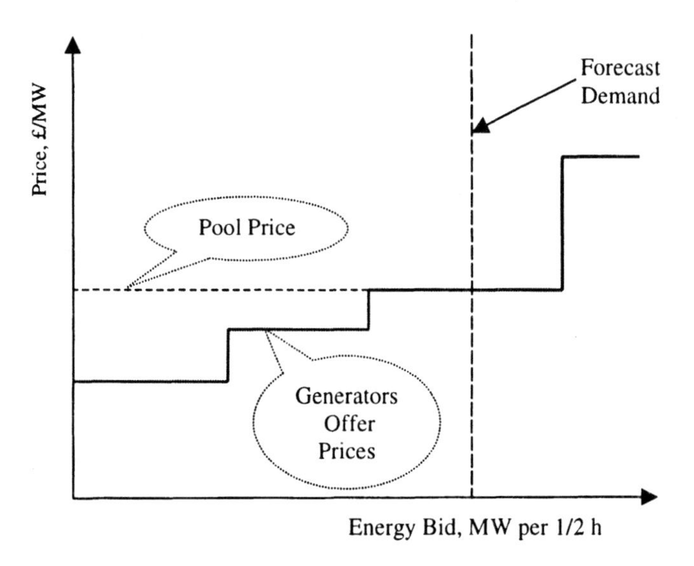

Figure 1. British power pool operation

Gencos sell power into the pool at the Pool Purchase Price (PPP) which is the SMP plus capacity payment, while the distribution companies and regional suppliers buy from the pool at the Pool Selling Price (PSP) which is PPP plus an *uplift*. The uplift is dependent on where the power is drawn, and takes into account the losses in network. In summary the following is the payment structure in the UK power pool:

Gencos receive: Pool Purchase Price (PPP)
= System Marginal Price + Capacity Payment

Customers (Regional Electricity Companies in this case) pay:
Pool Selling Price (PSP)
= PPP + Energy Uplift

2.1.2 The Nordic Electricity Market

In Chapter-1 we discussed the initiation of the Nordic deregulation process and the setting up of a common day-ahead spot-market. A distinctive feature of the Nordic electricity market is the significant share of bilateral transactions in total energy trades. The spot-market and other forms of market trades are conducted by NordPool, in which participation of generators or customers is not mandatory, unlike the UK pool. At present about 25% of the energy consumed in the Nordic countries is traded on NordPool. The volume is growing.

NordPool operates three types of markets, a futures market, a day-ahead spot market[2] and an hourly short-term market[3]. The futures market is not a physical market but purely a financial market for price hedging, risk management and trade in forward and future power contracts. Contracts for purchase of weekly base or peak load for up to three years in advance is possible. In the day-ahead spot market, bids are accepted for both sale and purchase, in the form of linear segments denoting price and quantity. The bids are aggregated into separate price versus quantity curves for supply and demand, which are matched to obtain the system price (*Figure 2*). The system price is applicable to all bidders- both buyers and sellers. In [5, 6 and 7] the structure and operation of Nordic market is discussed with reference to bid settlement and the price-area congestion management scheme in Norway. We shall discuss these issues in the context of transmission open access and congestion management in Chapter-4.

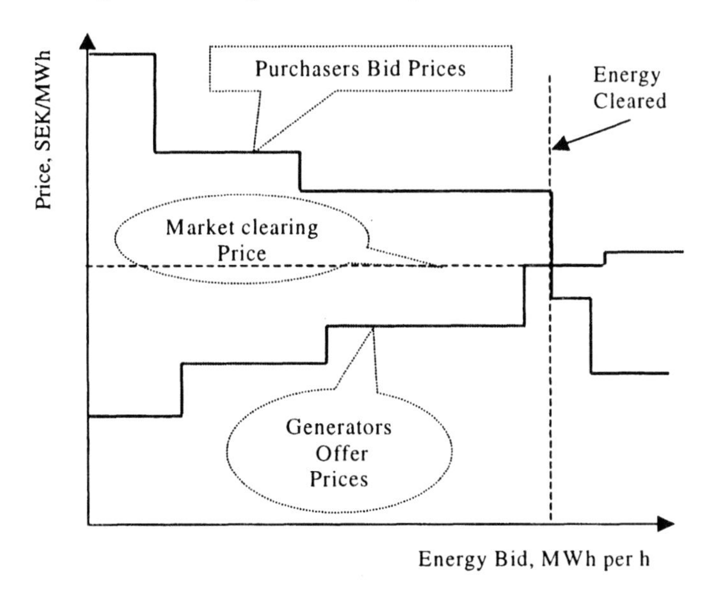

Figure 2. NordPool spot market settlement

Figure 3 shows the short-term market price in Sweden vis-à-vis the spot market price for a typical day while *Figure 4* shows the comparison on a day

[2] The day-ahead spot-market is known as *Elspot*.

[3] The hourly short-term market is known as *Elbas*. As of February 2001, it is working between Sweden and Finland only.

when there was an outage on one generator. The short-term market prices are evidently quite sensitive to system conditions, weather, etc. and there have been instances when these had shoot up drastically.

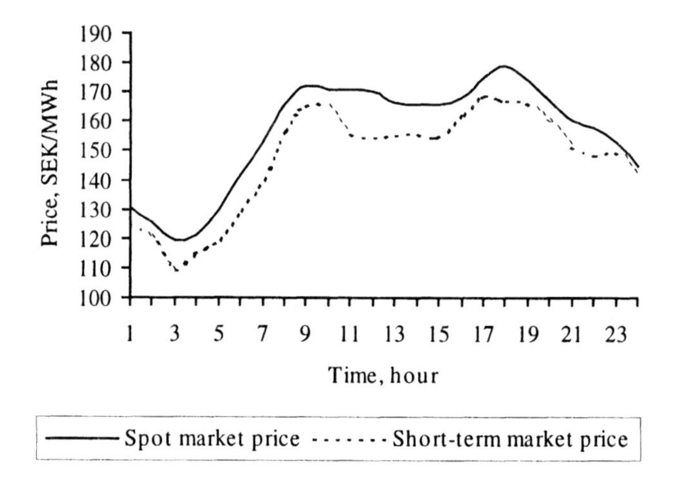

Figure 3. The short-term market price in Sweden (on an ordinary day- 20th November 2000) follows similar trends as the spot market price.

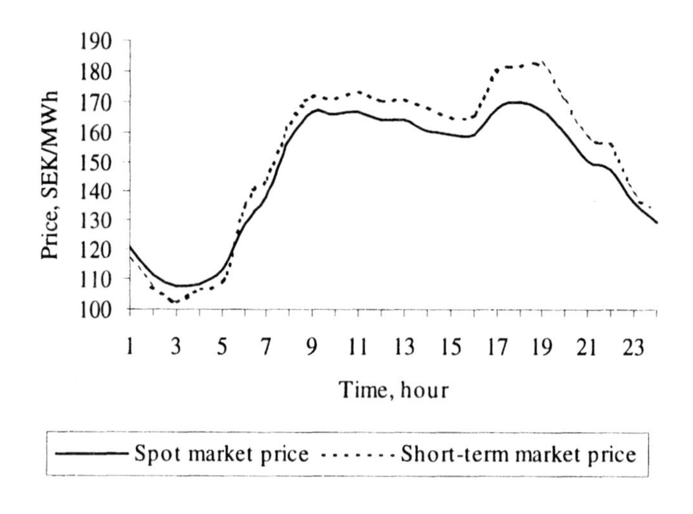

Figure 4. The short-term market price is sensitive to system conditions. On 7th November 2000, there was an outage of a 210 MW unit for 6 hours starting 10 AM and at 5 PM a 920 MW plant announced deferment of start-up from it maintenance-state by another few days. This case is not a major contingency on the system, but in times of more critical system conditions the short-term market prices can be very high.

2.1.2.1 The Energy Balance Market

In addition to the spot, short-term and futures markets being operated by NordPool, the national ISO of each country[4] operates its own regulating power market. Each ISO is responsible for minute-to-minute balance between production and consumption in its own area. To perform this task, a *balance market* has been established which comprises balance adjustment, balance regulation and balance settlement.

Since the ISO, as a grid company, does not own production or consumption capacity, such resources must be purchased in order to manage the momentary balance. Market participants who are able to regulate their generation or demand at short notice can place bids for upward or downward regulation. Bids are arranged in price order for each hour. In the operation phase the ISO uses the bids when a need for an adjustment of balance is necessary. The marginal price for the upward and downward regulation is registered for each hour and used for settlement of imbalances.

Figure 5 shows the time-scale of the operation of various markets in NordPool. It can be seen that the short-term market opens just two hours ahead of power delivery period. This market is meant for participants who have excess generation or customers who need excess energy at short notice. This market was initiated because of the long-time gap between the submission of bids to the spot-market, at 12:00 hours, to the actual power delivery time (00:00 to 24:00 hours of next day) which can be as high as thirty-six hours.

[4] Statnett in Norway, Svenska Kraftnät in Sweden, Fingrid in Finland, Elkraft and Eltra in Denmark are the respective ISO. Note that Denmark has a separate ISO for its Eastern and Western girds.

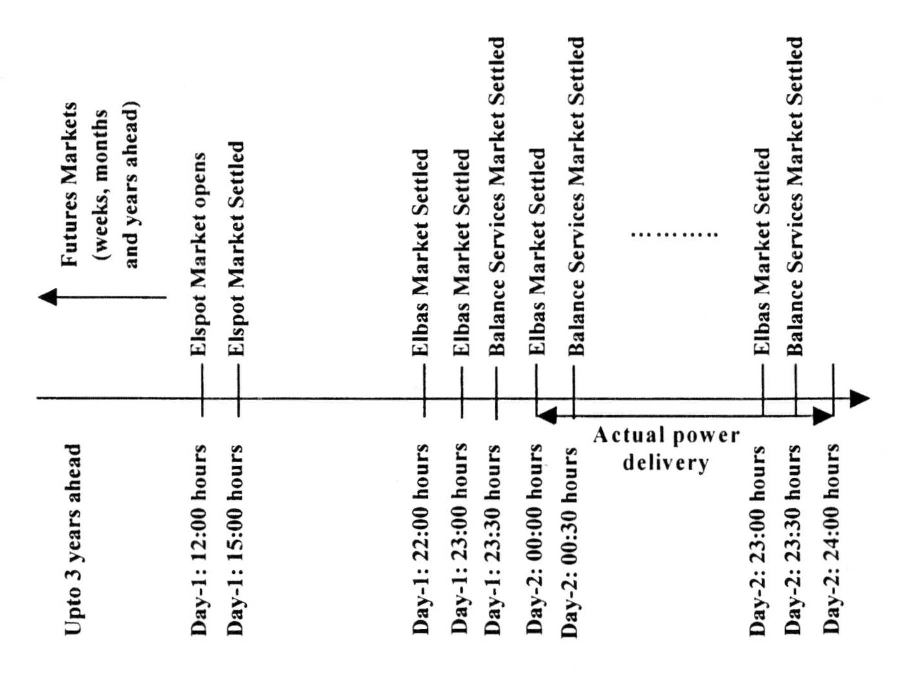

Figure 5. The Nordic electricity market on the time-domain

3. OPERATIONAL PLANNING ACTIVITIES OF ISO

3.1 The ISO in Pool Markets

In deregulated electric power systems operating with the pool type of structure, the ISO is responsible for carrying out market activities such as receiving bids from suppliers, unit commitment and dispatch simulations for the gencos and setting up the market clearing price. It also has the responsibility for ensuring system security and undertaking congestion management for which it has to procure ancillary services[5] and also decide on the various control actions to be taken.

[5] *Ancillary services* are all those activities necessary to support the transmission of power while maintaining reliable operation and ensuring the required degree of quality and safety. These would thus include regulation of frequency and tie line flows, voltage and reactive power control, system stability, maintenance of generation and transmission reserves, removal and/ or control of power harmonics and many others.

3.1.1 Activities of the ISO

On a time-scale starting from one day ahead of actual operation to real-time operation, the ISO's activities in a pool can be classified as follows:

24-hours ahead:

- Carries out a load forecast to determine the aggregate hourly load demand in its system the next day.

- Receives bids for supply of power from gencos. The bids can be of different types, and may sometimes also include information on unit commitment such as, start-up time, shut-down time, ramp rate limits and associated costs. Based on this information, it simulates the hourly dispatch using a model similar to the UC model discussed in Chapter-2 and evaluates the system price for each hour of the next day.

- Formulates the nodal marginal costs and congestion transmission prices based on the system price.

In real-time:

- Dispatches generation and load. Procures and provides for system services such as reactive power support, frequency regulation and other ancillary services.

After real-time:

- Calculates the settlements, nodal prices of energy and transmission congestion surcharges. Settlements include marginal fuel costs, capacity cost, start-up cost, congestion transmission charges, network services charges, and ancillary services payments.

The bids submitted by gencos can include different levels of information on their operations. For example, in the New York (NY) and PJM (Pennsylvania, Jersey, Maryland) power pools, the gencos bid their energy prices, start up costs, minimum and maximum generation levels [8]. In the case of PJM, a unit commitment model is simulated by the ISO to obtain the generation and commitment schedules. On the other hand, in NY-ISO it is assumed that the gencos have incorporated their unit commitment decisions while placing their bids. In this aspect, the NYISO is somewhat similar to the New Zealand Electricity Market (NZEM) where gencos have to carry out their own unit commitment and submit their ramping rates additionally with their energy bids [9]. The UK pool on the other hand, carries out a centralized unit commitment simulation based on the supply bids of gencos that include energy prices, ramp rates and other technical constraints. Some of the different types of generation scheduling practices adopted by various power pools are described in *Table 2*.

Table 2. Generation Scheduling Practices in Different Electricity Markets

Market	Information received by the market operator through bids	Who is responsible for Unit Commitment
UK	Energy price, start-up price, offered capacity, ramp up rate, ramp down rate, minimum generation level, minimum up time, minimum down time	ISO - the National Grid Company plc.
NY	Energy price, start-up costs, minimum generation level, maximum generation level	Self commitment by individual genco
PJM	Same as NY	PJM-ISO
NZEM	Energy prices, ramp rate limits, reserve bids	Self-commitment by individual gencos. The ramp rate limits submitted to the ISO are adhered to, in the bid clearing process
Nordic	Energy price and quantity for both buying and selling. There is also a short term market for trade close to real time	Self commitment by individual gencos

Two types of market settlement have been proposed for adoption by the market operator: (a) maximization of social welfare [9, 10, 11, 12, 13, 14, 15, 16 and 17] (b) minimization of consumer payment [18, 19, 20]. The first objective effectively minimizes the total costs due to the assumption that generation bids correspond to their actual costs. The later objective seeks to minimize the market-clearing price and hence the price the customers would pay for energy. Both approaches arrive at the same market conditions under certain conditions with certain simplifying assumptions.

3.1.2 Social Welfare Maximizing Market Settlement

Maximization of a social welfare objective function to obtain the optimum dispatch schedules has been the common practice in most centralized power pools.

Two cases may arise in this class of problems, one where the market operator receives both supply and demand bids and the system price is obtained by matching the highest priced sell bid to the lowest priced buy bid. The other is where only supply bids are received and the system price is obtained by finding the highest priced bid intersecting the system demand forecast. The former are called double auction pools and is used in New Zealand, California and NordPool markets while the later are called single auction pools and is typical of the UK pool.

3.1.2.1 Double Auction Power Pools

Let us consider markets where both, supply and demand bid is invited from participants. The system price is obtained by stacking the supply bids in increasing order of prices and the demand bids in decreasing order of their prices. The system price and the amount of energy cleared for trading is obtained from the crossing point of these curves as shown in *Figure 6*.

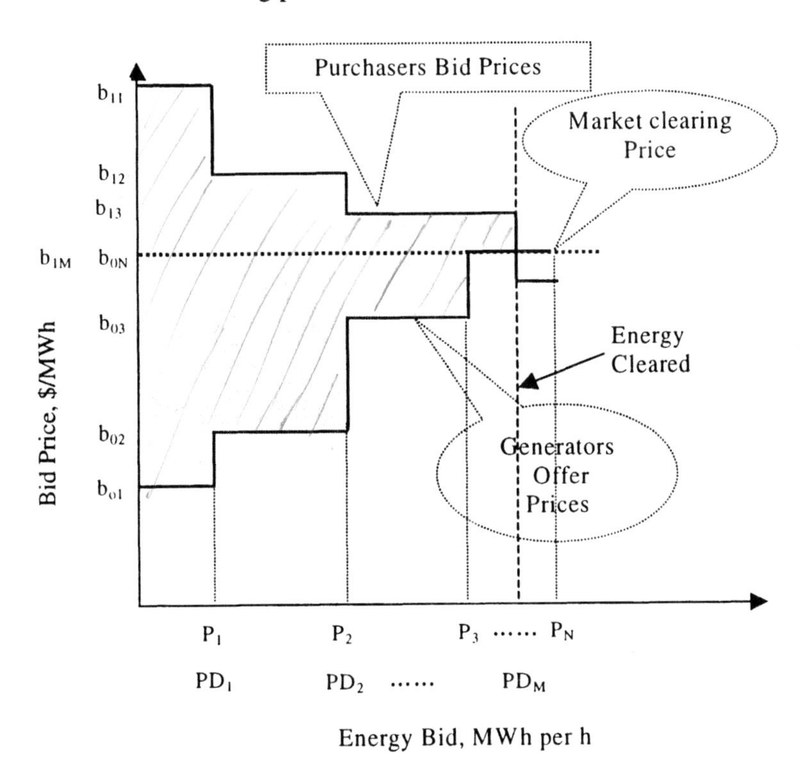

Figure 6. Market settlement in double auction power pools. The market operator seeks to maximize the social welfare function denoted by the shaded area in the figure.

Such auctions are called double auctions and *Figure 6* shows the typical market clearing process in such power pools. The shaded area in *Figure 6* denotes the social welfare from market based operation, that the market operator seeks to maximize. The bid prices in the figure show constant price bids, which need not be necessarily, so.

Let us assume that the market operator receives N supply bids with bid prices (BPS) from the participating gencos. Similarly, let the market operator also receive M demand bids from customers with bid prices (BPB). We can assume that BPS represents a genco's true marginal cost while BPB represents a customer's true benefit or utility function. We can thus write,

$$C_i = f(P_i); \qquad B_j = g(PD_j); \qquad \forall i \in N \text{ and } \forall j \in M \tag{1}$$

In (1), C_i is the cost function of genco *i* expressed as a function of power supplied P_i, while B_j is the benefit function of customer *j*, as a function of the power demand, PD_j. The social welfare function (J) can thus be defined as the total customer benefit, net of total cost to genco.

$$\text{Social Welfare, } J = \sum_{j=1}^{M} PD_j \cdot BPB_j - \sum_{i=1}^{N} P_i \cdot BPS_i \tag{2}$$

The first term in (2) denotes the total benefit or utility to the customers and the second term denotes the total cost to the gencos. Accordingly, the social welfare can also be written as follows:

$$J = \sum_{j=1}^{M} B_j - \sum_{i=1}^{N} C_i = \sum_{j=1}^{M} g_j(PD_j) - \sum_{i=1}^{N} f_i(P_i) \tag{3}$$

The social welfare function, given by (2) or (3), is maximized to obtain the market price. A demand-supply balance constraint, neglecting losses, need to be included, as follows,

$$\sum_{i=1}^{N} P_i = \sum_{j=1}^{M} PD_j \tag{4}$$

The supply bids may also include other parameters such as start-up price and ramp rate limits. Other constraints can also be appropriately included as per pool rules and settlement mechanisms. In the pool model of New Zealand [9, 10] the bids also specify different locations where the supply or demand will be made. Therefore, the market operator requires including transmission constraints in the settlement model. This is done using a dc power flow representation and the power dispatch adheres to the transmission limits.

Formulating the Lagrangian for the maximization problem described by (3) and (4), we have,

$$F = \sum_{j=1}^{M} g_j(PD_j) - \sum_{i=1}^{N} f_i(P_i) - \lambda\left[\sum_{j=1}^{M} PD_j - \sum_{i=1}^{N} P_i\right] \tag{5}$$

Applying the *Kuhn-Tuckers condition of optimality*, similar to that discussed in Chapter-2, we can write:

$$\frac{dg_j(PD_j)}{dPD_j} = \lambda$$
$$\frac{df(P_i)}{dP_i} = \lambda \tag{6}$$

The Lagrange multiplier λ denotes the system marginal price, which stipulates a rational participant to set its price equal to the marginal utility.

3.1.2.2 Single Auction Power Pools

This section discusses the market settlement mechanism where only gencos submit their bids that are then stacked in the increasing order of prices. The highest priced bid to intersect with the system demand forecast determines the market price. This arrangement is found in the UK, wherein the load forecast for every half-hour determines how much quantity (in MWh) of supply bids have to be cleared by the pool (*Figure 7*).

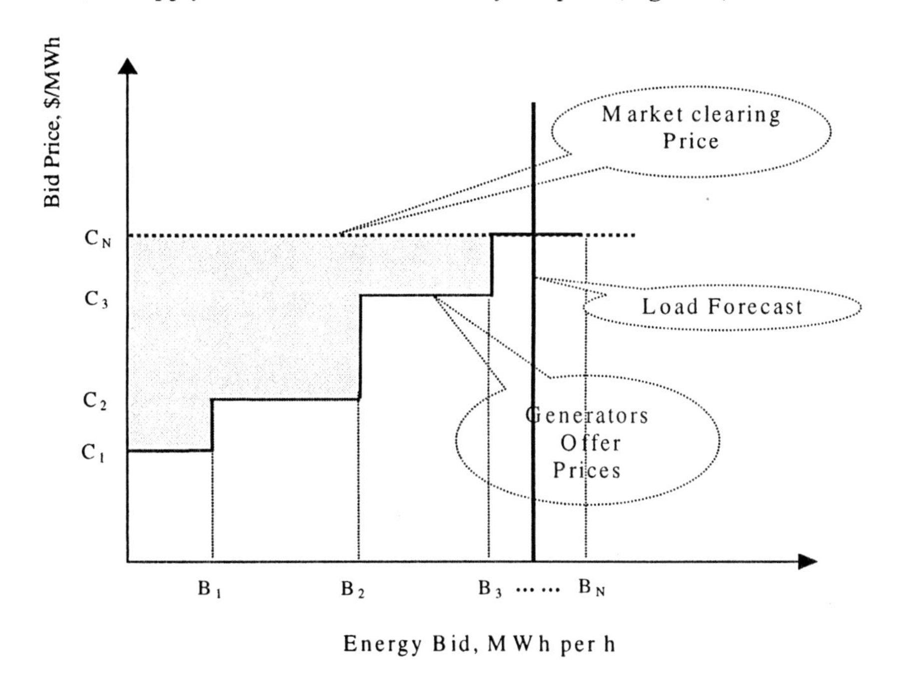

Figure 7. Market settlement in single auction power pools. The market operator seeks to minimize the system cost (which is effectively maximization of the social welfare, denoted by the shaded area in the figure).

In this case the social welfare maximization problem is similar to minimization of the gencos' costs (assuming that gencos' bid prices are their true marginal costs) and a scheduling model can be defined as follows:

$$\text{Minimize, } J = \sum_{i=1}^{N} BPS_i \cdot P_i = \sum_{i=1}^{N} f_i(P_i) \tag{7}$$

The above objective function is minimized subject to the demand-supply balance constraint, neglecting losses,

$$\sum_{i=1}^{N} P_i - PD = 0 \tag{8}$$

We can again formulate the Lagrangian as follows,

$$F = \sum_{i=1}^{N} f_i(P_i) - \lambda\left(\sum_{i=1}^{N} P_i - PD\right) \tag{9}$$

and the *Kuhn-Tuckers condition of optimality* is,

$$\frac{dC_i(P_i)}{dP_i} = \lambda \tag{10}$$

In the UK power pool the market settlement process also incorporates unit commitment decisions and therefore the gencos are also required to submit a start-up price bid, in addition to the variable component. Accordingly, the unit commitment decisions are incorporated in the social welfare objective function as follows:

$$J = \sum_{k=1}^{T} \sum_{i=1}^{N} BPS_{i,k} \cdot P_{i,k} + ST_i \cdot UST_{i,k} \tag{11}$$

In (11), the index k denotes the bid time period, which is half an hour in UK, and T stands for the entire market scheduling horizon, *i.e.* 48 time intervals for the next day. UST is a binary variable denoting the unit start-up decision, while ST represents the start-up price offer [14]. The above objective is minimized subject to the constraints on demand and reserve requirements, as well as ramp rate limits.

3.1.2.3 Example

(a) Double Auction Market

Consider a double auction market wherein the market operator receives supply-bid offers from five gencos (i=5) and demand-bid offers from eight customers (j=8) as given in *Table 3*.

Table 3. Example bid offers received by market operator

	Base Supply Bid Offers						Base Demand Bid Offers	
Offer	**ST, \$/hr**	**BPS, \$/MWh**	**P^{Max}, MW**	**P^{Min}, MW**	**RUP, % of P**	**RDN % of P**	**BPB, \$/MWh**	**P^{Max}, MWh**
1	1000	20.50	750	150	125	75	32.35	250
2	1500	21.50	600	100	135	65	30.59	350
3	750	25.55	200	50	145	55	28.25	500
4	500	22.00	400	100	125	75	27.75	100
5	100	27.50	150	40	155	45	23.50	200
6	-	-	-	-	-	-	22.60	350
7	-	-	-	-	-	-	20.30	400

Based on the above offers the market operator settles the market by maximizing a social welfare function, given by,

$$J = \sum_{k=1}^{24} \sum_{j=1}^{8} BPB_{j,k} \cdot PD_{j,k} - \left(\sum_{k=1}^{24} \sum_{i=1}^{5} BPS_{i,k} \cdot PG_{i,k} + ST_{i,k} \cdot UST_{i,k} \right) \tag{12}$$

In the above, PG is the generation cleared in the market over and above the minimum offer, P^{Min}. The demand-supply balance can be written as,

$$\sum_{i=1}^{5} W1_{i,k} \cdot P_{i,k}^{Min} + PG_{i,k} = \sum_{j=1}^{8} PD_{j,k} \tag{13}$$

Note that the dual of the above demand-supply constraint gives the market price. W_1 is a binary variable, denoting clearance or otherwise, of a generator's bid offer while W_2 is a binary variable denoting, the clearance or otherwise, of a customers demand offer. The upper limits on clearance of the generation and demand bids are given as follows,

$$\sum_{i=1}^{5} PG_{i,k} \le W_{1_{i,k}} \cdot \left(P_{i,k}^{Max} - P_{i,k}^{Min} \right) \qquad \sum_{j=1}^{8} PD_{j,k} \le W_{2_{j,k}} \cdot P_{j,k}^{Max} \tag{14}$$

The ramp-rates quoted by the gencos also need to be taken into account in the market settlement process, as follows:

$$\begin{aligned} P_{i,k}^{Min} + PG_{i,k} &\le RUP_{i,k} \cdot \left(P_{i,k-1}^{Min} + PG_{i,k-1} \right) && \forall\ k > 1 \\ P_{i,k}^{Min} + PG_{i,k} &\ge RDN_{i,k} \cdot \left(P_{i,k-1}^{Min} + PG_{i,k-1} \right) && \forall\ k > 1 \end{aligned} \tag{15}$$

We show the results of the market settlement for a single bidding period in *Table 4*. In such as case however, the ramp rate limits and minimum quantity offers need not be considered.

Table 4. Market clearing for double auction, single hour

Selected Offers	Generation cleared, MW	Demand Cleared, MW	System Price, $/MWh
1	750	250	
2	Not cleared	350	
3	Not cleared	500	27.50
4	400	50	
5	Not cleared	Not cleared	
6	-	Not cleared	
7	-	Not cleared	
8		Not cleared	

From *Table 4* we note that, although the bid-price offer of genco-2 is less than that of genco-4, its offer is not cleared in the market because of its significantly higher start-up price. The system price is the price where the aggregated supply curve intersects the aggregated demand curve, *i.e.* at the bid-price of customer-4, which is the bid cleared at the margin. A total of 1150 MW has been cleared in the market that is being supplied by two gencos and being purchased by four customers.

It is to be noted that all parameters offered through the bids could vary for each bidding period. Thus, the minimum generation offered (P^{Min}), the maximum generation or demand offered (P^{Max}), the ramp rate quotations (RUP and RDN), the price offers for energy (BPS and BPB) and start-up price offer (ST) for each participant, would be time-dependent. In such a case, an elaborate model solution as described by (12)-(15) will be necessary.

(b) Single auction Market

From the viewpoint of a mathematical model formulation, there is not much difference between the double auction market and the single auction market. The social welfare objective function can now be written as,

$$J = \sum_{k=1}^{24} \sum_{j=1}^{8} BPS_{i,k} \cdot PG_{i,k} + ST_{i,k} \cdot UST_{i,k}$$

The demand-supply balance equation is now of the form,

$$\sum_{i=1}^{5} W1_{i,k} \cdot PMin_{i,k} + PG_{i,k} = PD_k$$

Note that in this case, the demand PD is a pre-determined quantity available with the market operator before settlement. Other constraints remain the same as discussed with reference to the double auction model.

Assuming that the single auction market is also required to clear a total demand of 1150 MW, *i.e.* the amount cleared in the double-auction market, we get the same solution, *i.e.* genco-1 sells 750 MW and genco-4 sells 400 MW, while the other gencos' bids are not cleared.

However, the market-clearing price is now determined one-sided, *i.e.*, only from the price offer of the genco that is cleared at the margin. Hence, the market-clearing price now becomes,

$$\text{Market clearing price} = \frac{\$500/\text{hr} + \$22.0/\text{MWh} \cdot 400\text{MWh}}{400\text{MWH}} = \$23.25/\text{MWh}$$

(c) Minimization of Consumer Payment

Alongside the social welfare maximization based pool settlement philosophy, another school of thought has emerged that the conventional methods of evolving the pool prices, based on maximization of net social welfare, discussed above, are not efficient enough [18, 19, 20]. It has been argued that these methods often fail to capture the complexities of the markets and would not necessarily minimize the cost to the customer, particularly so when the system price is evaluated taking into consideration the no-load and start-up costs. The offered alternative is to minimize the system price that the customers finally pay for their energy.

This is explained using a simple example from [19] as follows. Consider that three generating units with their corresponding bid price offers, denoted by BC_i, as given below, have to be dispatched by the market operator to meet a forecasted load of 800 MW.

$$BPS_1 = 1000\$/\text{hr} + 20 \cdot P_1 \$/\text{MWh} \qquad PMax_1 = 750\text{MW}$$

$$BPS_2 = 1000\$/\text{hr} + 20 \cdot P_2 \$/\text{MWh} \qquad PMax_2 = 750\text{MW}$$

$$BPS_3 = 100\ \$/\text{hr} + 24 \cdot P_3 \$/\text{MWh} \qquad PMax_3 = 150\text{MW}$$

In the above bid prices, the first component denotes the unit start-up and no-load price bid while the second component is the unit variable price bid which depend upon a unit's generation. The pool price will be obtained from the price of the highest accepted bid for dispatch.

With the least-cost objective (assuming that the bid prices represent costs), the market operator will obtain the following unit commitment and dispatch decisions:

P^*_1 = 750 MW, P^*_2 = 0, P^*_3 = 50 MW. The corresponding system price, which is maximum of the two genco prices, inclusive of start-up and no-load prices, will be,

$$\text{System Price} = \max\left\{\frac{1000+20\cdot 750}{750}, \frac{100+24\cdot 50}{50}\right\} = \max\{21.33, 26.00\}$$

Therefore, the system price in the least cost dispatch will result in system price of $26.0/MWh, as explained in *Figure 8*.

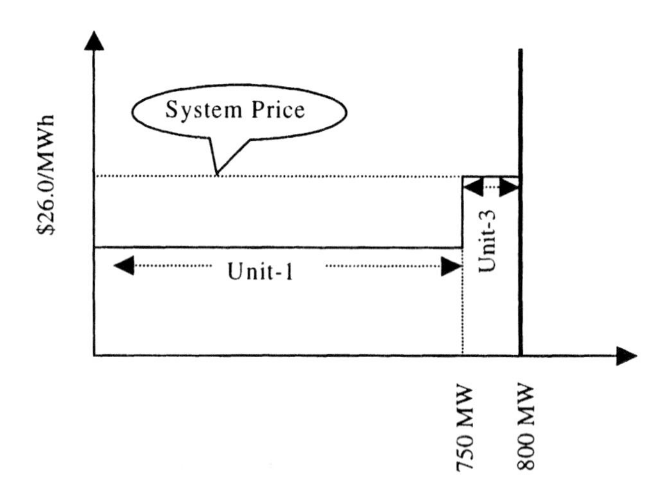

Figure 8. Least cost dispatch

With the consumer payment minimization objective the generation schedule is obtained as follows:

P^*_1 = 400 MW, P^*_2 = 400, P^*_3 = 0 MW. Again, the system price can be calculated as follows:

$$\text{System Price} = \max\left\{\frac{1000+20\cdot 400}{400}, \frac{1000+20\cdot 400}{400}\right\} = \max\{22.50, 22.50\}$$

In this case, the system price is $22.50/MWh and hence the consumer payment is lesser (*Figure 9*). The total system cost, though, is greater compared to the cost minimization case.

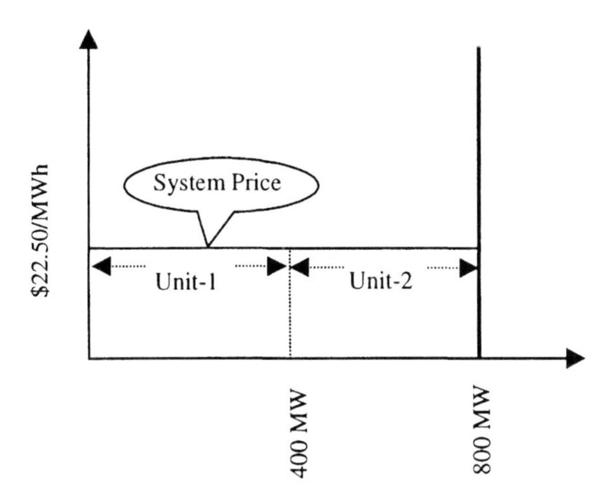

Figure 9. Consumer cost minimizing dispatch

3.2 The ISO in Bilateral Markets

In bilateral contract markets, the role of the ISO is limited to system security management, congestion management and reliability aspects. To this end, the ISO has to procure various ancillary services from ancillary service providers. The ISO's function in a bilateral contract market can also be described on the following time-scale, from one day ahead of actual operation to real-time.

24-hours ahead:

- The ISO is informed of all the hour-by-hour transactions that are to take place the next day. These transactions can either be decided by an independent market operator based on a bidding mechanism, as in the hourly spot market in NordPool, or through bilateral contract between the genco and the customer, based on independent negotiations. The ISO is not involved in any of these processes.

- Once the transactions are available to the ISO, it carries out power flow studies and other simulations based on its load forecast for the next day, availability of transmission capacity, and other factors to determine the level of system security. If required, it makes provision for additional transmission capacity, ancillary services, or orders curtailment of certain transactions.

In real-time:

- Monitors the system for power flows, frequency and voltage conditions and trades. Interacts with the regional networks and control rooms for providing adequate frequency regulation services, reactive support and voltage control services

- Power imbalances are corrected by making provision for parties, which shall be responsible for this and charge the same to the defaulting parties.

After real-time:

- Settlement of the accounts and payments for the ancillary service providers.

The role of the ISO is much less in such an open access environment. The functions are limited to verifying the feasibility of trades 24 hours ahead; monitor them in real time for system security maintenance and provide for ancillary services.

3.2.1 Analysis of Bilateral Markets

Bilateral transactions are contracts between sellers (can be gencos, brokers or their agents) and buyers (can be discos, large customers, brokers or their agents) and can take place in numerous forms. The transactions can be firm (non-curtailable) or non-firm (curtailable), short-term or long-term, for energy, instantaneous power or reserve. With time, additional innovative types of transactions are expected to emerge, limited only by the ingenuity of the market and the likelihood of making a profit [21].

There is no role of the ISO in formation of these contracts and the two transacting parties are free to negotiate their prices. However, once the transactions are negotiated, the ISO needs to be informed about the trades since it is responsible to ensure that the transactions do not endanger system security.

Modeling of bilateral transactions is usually through the use of bilateral transaction matrices (BTM) as proposed in [21 and 22], denoting the linkages between various parties involved. The construction of BTM requires adhering to certain rules, such as the column rule and row rule [23] which ensures that the BTM elements satisfy the basic constraints of demand supply balance. Assuming that a buyer and seller can be located at the same bus, we can formulate the following relations for a BTM:

Consider a system of n-buses and bilateral transactions can take place between any generator and any load, located at any bus. It may also take place between a generator and a load located at the same bus. The BTM, denoted by T, can then be defined as follows:

$$T = \begin{bmatrix} T_{1,1} & T_{1,2} & \cdots\cdots & T_{1,n} \\ T_{2,1} & T_{2,2} & \cdots\cdots & T_{2,n} \\ \cdots & \cdots & \cdots\cdots & \cdots \\ T_{n,1} & T_{n,2} & \cdots\cdots & T_{n,n} \end{bmatrix} \quad (16)$$

In (16) the elements $T_{i,j}$ over a row define the bilateral contract of a generator *i* with all possible loads *j*. Thus the first row of T ($T_{1,j}$) defines the bilateral contracts a generator at bus-1 may enter into with loads at buses *j*.

Similarly, the column elements define the bilateral contract a load *j* may enter into with all possible generators *i*. Thus the first column of T ($T_{i,1}$) defines the bilateral contracts a load at bus-1 enters into with other generators at bus *i*. Accordingly, the matrix T requires to fulfill the following conditions for feasibility of the bilateral transactions.

$$\begin{aligned} &\sum_i T_{i,j} = PD_j \\ &P_i^{Min} \le \sum_j T_{i,j} = P_i^{SCH} \le P_i^{Max} \end{aligned} \quad (17)$$

In (17), PD is the demand at bus *j* while P^{SCH} is the generation scheduled by a genco located at a bus *i* to meet its bilateral transaction commitments. This is submitted to the ISO in advance. Understandably, the generation scheduled, should be within the upper and lower bounds of generation from the unit, P^{Max} and P^{Min} respectively.

From the perspective of the ISO in bilateral markets, its objective is to ensure that the system is secure and reliable. Therefore, under certain circumstances, it might be necessary for the ISO to curtail some of the transactions for system security reasons. The choice of curtailment of transaction is important to the parties involved in them, since curtailment would affect the financial deals. Therefore, the ISO should act in an impartial and fair manner to all parties, while deciding on the curtailment of transactions.

The minimization of an objective function comprising the deviation from scheduled transaction is an option for the ISO. The objective can be expressed as the sum of the squares of deviation of actual allowable transactions from contracted transactions.

$$\min \ TRDIFF = \sum_{i,j} (T_{i,j} - T_{i,j}^{Allowable})^2 \quad (18)$$

In (18), T is known *a priori* while $T^{Allowable}$ is the allowable transaction decision variable determined from the ISO's optimization model. TRDIFF is the objective function that the ISO seeks to minimize.

The bilateral transactions need to be included in the system real and reactive power flow equations (discussed in Chapter-2), as follows:

$$P_i^{SCH} - \sum_j T_{i,j}^{Allowable} = \sum_j |V_i||V_j| Y_{i,j} \cos(\theta_{i,j} + \delta_j - \delta_i) \tag{19}$$

$$Q_i - QD_i = -\sum_j |V_i||V_j| Y_{ij} \sin(\theta_{ij} + \delta_j - \delta_i) \tag{20}$$

V is the bus voltage, δ is the angle associated with V, $Y_{i,j}$ is the element of bus admittance matrix and θ is the angle associated with $Y_{i,j}$.

Since bilateral transactions usually do not specify anything about the reactive power associated with the transactions, it is the responsibility of the ISO to provide for adequate reactive power support. Thus Q_i in (20) is the reactive power schedule provided by the ISO which includes switching decisions of shunt capacitors and also reactive power support from generators. QD is the ISO's estimate of reactive power demand in the system for the bilateral transactions scheduled.

Other constraints should also be included in the ISO's optimization framework, as described below:

- Upper and lower limits on bus voltages
- Upper and lower limits on bus reactive power support
- Upper and lower limits on generator reactive power capability
- Transmission limits on the lines

In the above model, it was stated that the ISO's objective is to minimize TRDIFF. However, such an objective function may not be able to achieve the most desirable solution, as explained below.

For example, in order to seek a least curtailment solution, the ISO might have to procure excessive reactive power support at some buses (including that from generators). The least-curtailment solution may also result in increased system losses, thereby requiring the ISO to procure increased loss compensation service from the ancillary service market.

Under both the circumstances the ISO has to buy services- reactive power support or loss compensation, in excess of its requirements, so as to minimize the transaction curtailment. This is an undesirable consequence from the perspective of the ISO.

These are conflicting objectives between themselves and minimization of one can lead to an increase in the other. The ISO may however, choose to attach a priority order to them and obtain the optimal decision in a multi-objective framework. Combining the conflicting objectives using a compromise function is another alternative to obtain an optimal solution [24]. The compromise function, given by (21), represents the objective of the ISO of meeting the contradictory requirements simultaneously.

$$J = \sqrt{\left(\frac{\text{TRDIFF}}{\text{TRDIFF}^*}\right)^2 + \left(\frac{\text{Loss}}{\text{Loss}^*}\right)^2 + \left(\frac{\text{Payment}}{\text{Payment}^*}\right)^2} \tag{21}$$

In (21), TRDIFF is the transaction difference that the ISO seeks to minimize, Loss is the total transmission loss that the ISO has to compensate from other supplies, and Payment is the total money paid out by the ISO to its various service providers. Note that these are decision variables that the ISO seeks to simultaneously minimize. TRDIFF*, Loss* and Payment* are the corresponding minimum transaction difference, minimum loss and minimum payment, when these are minimized individually.

In [11] an objective function based on how much a customer is willing to pay so that its transaction is not curtailed, has been used. The ISO can also use such an objective for minimization that will determine curtailment of those transactions where the parties are not willing to pay.

4. OPERATIONAL PLANNING ACTIVITIES OF A GENCO

In deregulated electricity markets, the genco is usually an entity owning generating resources and participating in the market with the sole objective of maximizing its profit, without concern for the system unless there is an incentive for it. Therefore, the genco's operational planning activities differ with the market structure in which it is operating.

4.1 The Genco in Pool Markets

In pool markets the gencos are required to bid for energy supply and associated price to the market operator (usually, also the ISO). In some markets the gencos are also required to provide start-up price offers, ramp rates, minimum up and down time, *etc.* (*refer Table 2*). The market operator in turn carries out a unit commitment to obtain the optimal dispatch and settles the market and formulates the generation schedules.

However, in certain pool markets, the market operator does not carry out the unit commitment and the gencos therefore, do not have to submit any other information apart from their bid prices and quantities. The operational planning activities of a genco operating in a pool market can be outlined on the time domain as follows:

24-hours ahead:

- A price forecast is available or carried out to estimate the hourly market prices for the next day

- Based on its generating unit characteristics, unit availability, ramp rate, etc., determines a bidding strategy for each bidding period next day

In real-time:

- Meet generation schedules as ordered by the ISO

4.2 The Genco in Bilateral Markets

In bilateral markets, the gencos enter into direct buy and sell contracts with selling / buying agents. These contracts are usually for trade in energy supply but it could also include contracts for reserves or other services. The contracts are usually on a long-term basis and the prices are decided *a priori*, though short-term bilateral contracts do also take place.

Further, if in such systems, a spot market for energy trading on an hour-to-hour basis exists, the genco can sell its excess generation to the market and harness additional profit. However, for participating in the market as a seller or as a buyer, the genco has to submit bids for price and quantity, or any other parameter as required by the market design, usually one day in advance. As we discussed earlier, this is the existing structure in the Nordic countries, and the gencos are themselves responsible for their unit commitment while they also participate in spot-markets where the bids are for price and quantity.

Therefore, all the technical constraints such as ramp rate limits and start-up costs need to be internalized by the genco in its bid price. It also involves assumptions on generation dispatch, which should be realistic enough in order to avoid financial losses in the spot market. The problem is more complex if the genco has a predominantly thermal generating capacity thereby reducing its flexibility of operation as against those with predominantly hydro capacity. The operational planning activities of a genco operating in a bilateral market can be outlined on the time domain as follows:

24-hours ahead:

- A price forecast is available or carried out to estimate the hourly market prices for the next day

- Using the price forecast as input, the genco determines its unit commitment, generation schedule and trading decisions for the next day so as to meet its bilateral contract commitments and maximize its profit from the market.

- Based on its trading decisions and price forecast it determines the bidding strategy for each bidding period during the next day

- Submits the decided dispatch schedule to the ISO

At time periods near real-time:

- Monitors the system conditions, monitors balance market prices and obtains updated forecast of the balance market prices.

- Formulate bidding strategies for the balance market as appropriate

In real real-time:

- Fulfils generation commitments to bilateral contract customers and day-ahead market commitments

- Fulfil the balance market commitment if bid is selected

4.3 Market Participation Issues

There are various complex issues arising from uncertainties in market prices and technical constraints associated with unit operations, which need to be considered by the genco while scheduling its generating units and optimizing its generation so as to meet the bilateral contracts and trade during the next day.

In addition to finding the dispatch and unit commitment decisions while maximizing profit, its scheduling models should include trading decisions such as market buy and sell. Subsequently, the problem of finding appropriate bidding strategies for the trading decisions emerges.

A desirable bidding strategy should consider the complex inter-play between technical aspects of unit operation such as ramping, limits, and minimum run-time, economic interests of the genco as well as the uncertainties associated with market trading.

Figure 10 shows the operations planning activities of a genco in bilateral markets with the option of participating in the daily spot market. This is also applicable to gencos in pool markets where unit commitment decisions have to be internalized while placing their bids.

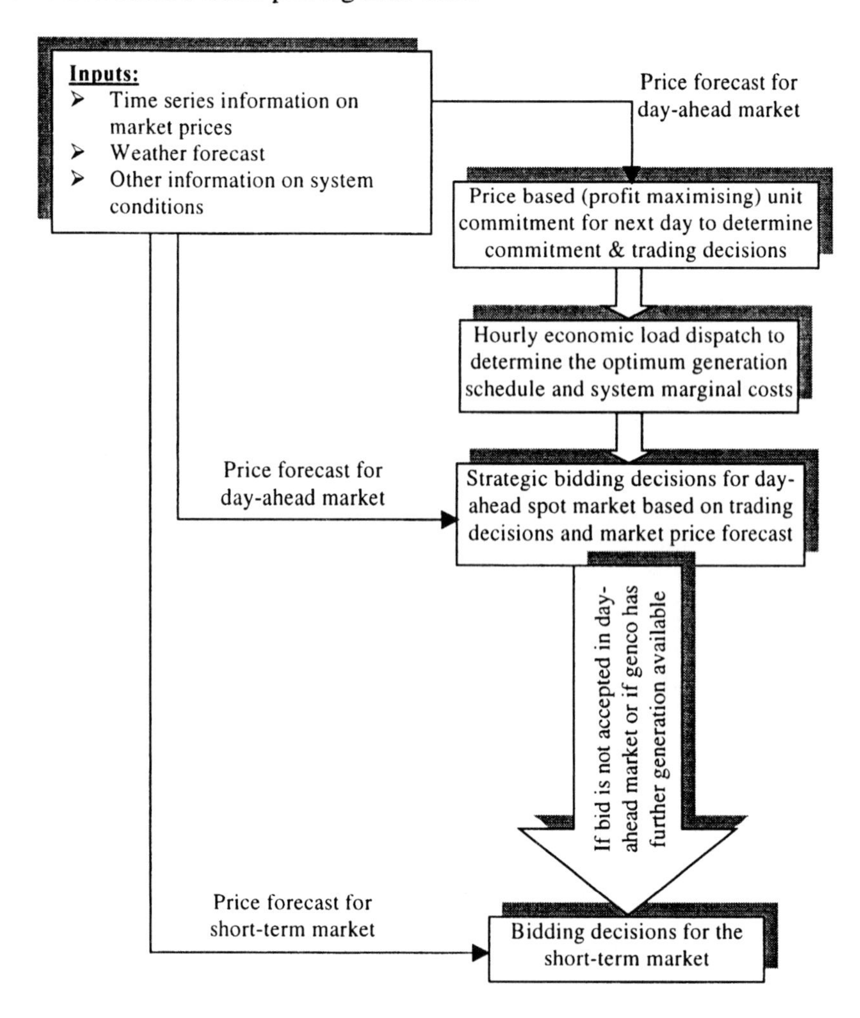

Figure 10. Short-term operational planning activities of a genco participating in electricity market trades

Depending on the spot market price forecast model in use, the input to the model would vary. However, the basic inputs to the price-forecast model are the past time-series data of spot market prices and the next day's weather forecast data. Using this information, the hourly price forecast is obtained which is then fed into the genco unit commitment module. The unit commitment module, which will be discussed later, is specific to the genco and includes the objectives and interests particular to the genco. Once the next day's unit commitment and trading decisions for buy / sell are obtained, the next stage is to determine the bidding strategies. The bidding strategies are formulated using the inputs such as- marginal cost at the evaluated operating schedule and price forecast information.

In markets where there is a provision for short-term trading, the gencos need to forecast the short-term market prices also and formulate its short-term bids on an hour to hour basis.

4.3.1 Market Price Forecast

An important input to the decision-making activities of a genco is a good forecast of the market prices. In the context of this chapter, price forecasting refers to day ahead or a week ahead forecast of the price- particularly the price in the spot market or the short-term market. This is important because an accurate forecast of the short-term market price helps the genco to bid for power sell or buy appropriately and strategically, thereby providing higher returns. Bilateral contract prices also have a tendency to be indirectly affected by spot-price trends. Thus good spot market price forecasts can help set up profitable bilateral contracts. In the short-term markets, such as that in Sweden and Finland, discussed in Section-2.1.2 of this chapter, continuous trading up to two hours in advance of real time is possible. In these markets, the prices can be highly volatile to system conditions such as sudden outages, and external factors such as temperature variations, rainfall, etc. It is usually of great interest to gencos and other market players to have a good forecast toolbox for these prices.

Price forecast in the general sense can also include forecast of futures and forward market prices. These forecasts may be carried out months or even a year in advance. These forecasts may be useful if the genco is contemplating investments in generation capacity, market risk analysis, production and maintenance planning, among others.

Most often the genco has an in-house price forecast tool based on available forecasting methods such as the conventional linear regression analysis technique, to cater to the need of a price forecast.

Two basic approaches to price forecasting have been discussed in [25]; one is the statistical method that uses time-series analysis of past data of market prices to develop a model which can fairly closely forecast the price for the next day. The main problem in this approach is to estimate the parameters of a mathematical model based on historical data. The other method of price forecast is by simulation, which incorporates the generation dispatch patterns over a period of time.

A stochastic model of the form given in (22) is used in [26] to determine the price forecast in the Norwegian electricity market.

$$\rho_n(k+1) = a_p(k) \cdot \rho_n(k) + \in(k) \tag{22}$$

$\rho_n(k)$ is the normalized average price in the week, a_p is the auto-correlation coefficient determined from price forecasts and ε is a normal random variable and k is the index of time variable.

Another approach recently developed is the application of artificial neural networks to short-term price forecasting [27], making use of the system demand forecast and system reserves information as inputs to the model.

4.4 Unit Commitment in Deregulated Environment

This section discusses a typical unit commitment model that may be used by a genco operating in bilateral markets with unit commitment constraints internalized in the bids. In a competitive environment, the genco in principle has no other objective but to produce electricity and sell it with maximum profit. This can be re-stated as the following UC problem formulation:

Given a forecast of spot market prices, establish a generation schedule that maximizes the profit over the planning period, subject to meeting all relevant constraints.

As long as the objective is profit, the UC scheduling can be done as if all generation, after fulfilling the bilateral contracts, is sold in the spot market. The scheduling models are generally based on a *price-taking* assumption, which means that no participant can influence the spot market price single-handedly. This is an ideal condition and the market in such a case is known as a *perfectly competitive market.* The usually large number of market participants makes a price-taking assumption reasonable.

The genco operating in such markets, often have a mix of customers- a part of their load is under direct bilateral contracts and their prices are fixed *a priori*. Secondly, the genco has an unrestricted option to participate in the spot-market as a buyer and/or as a seller. It also needs to maintain a certain amount of reserve generation available at all hours to account for contingencies, load variation or spot-market price fluctuations.

In this framework the genco has to devise its operating plan for the next day and take decisions on the following-

- The unit commitment schedule
- The trading level (buy/sell schedule) in spot market
- The operating strategy to choose

It can be safely assumed that the genco is equipped with an hourly spot market price forecast for the next day in order to formulate its strategies. In the following sub-sections we discuss a typical UC model applicable to a genco in deregulated markets seeking to maximize its profit [28, 29].

4.4.1 Objective Function of the Genco

The objective function of the genco in such a market can one of the many possible, but most importantly, it would seek to maximize its profits. The profit over a day for the genco can be expressed as:

$$\text{Profit} = \text{Revenue from}\left[\text{Spot market sell} + \text{bilateral power sell}\right] - \text{Payment for}\begin{bmatrix}\text{Spot market buy} + \text{unit operating costs} + \\ \text{start-up costs} + \text{shut down costs}\end{bmatrix} \quad (23)$$

This can be mathematically expressed as,

$$\text{Profit} = \sum_{k}\begin{bmatrix}\rho_{Mk}\cdot PSell_k + BC_k\cdot CP_k - \rho_{Mk}\cdot PBuy_k \\ -\sum_{i=1}^{NG}\left\{W_{i,k}\cdot CMin_i + PG_{i,k}\cdot GCst_i + UST_{i,k}\cdot ST_i + USD_{i,k}\cdot SD_i\right\}\end{bmatrix} \quad (24)$$

In (24), k is an index of time, ρ_M is the spot-market price, PSell and PBuy are decision variables denoting the amount of power to be traded (sold and purchased, respectively) from the spot-market. BC is the bilaterally contracted power at a price CP. Note that the component of revenue from bilateral transactions is a constant since bilateral contract prices are fixed *a priori*. Though it will not affect the optimum solution the term is included to help understand the different components of a genco's financial transactions. CMin is the generation cost at minimum generation limit of the unit (P^{Min}), GCst is the generation cost beyond PMin, ST is the unit start-up cost and SD is the unit shut-down cost. W, UST and USD are binary variables denoting unit status (1 = ON, 0 = OFF), unit start-up status (1 = Start-up, 0 = NO) and unit shut-down status (1 = Shut-down, 0 = NO) respectively.

4.4.2 Constraints

4.4.2.1 Demand-Supply Balance

Unlike the conventional UC discussed in Chapter-2 where the generation was scheduled to match the total demand plus a certain amount of reserve, the genco in a deregulated market is now no longer responsible for matching the scheduled generation to the demand. It is in fact only seeking a part of the whole system demand for itself.

The genco would thus seek to schedule that much generation that provides it with the maximum profit. Therefore, the demand-supply balance in deregulated UC is somewhat of an open-ended constraint with no particular requirement of meeting the demand.

The 'so-called' demand-supply balance would seek to match the generation and spot market purchase by the genco with its total bilateral contracted demand and the spot market sell after providing for adequate reserves for itself. Also note that in such models, the provision for simultaneous buying and selling is also possible. Therefore, the new demand-supply balance of the genco operating in a deregulated market is given as,

$$W_{i,k} \cdot P_i^{Min} + PG_{i,k} + PBuy_k - PSell_k = BC_k + RESV_k \tag{25}$$

RESV is the reserve generation capacity that the genco may choose to keep available in order to cater to demand changes, or price fluctuations in the short-term market.

4.4.2.2 Limit on Power Sell

The genco should also ensure that the spot market buying and selling schedule that evolves from the UC solution is met. This means that the genco should be able to meet its energy transaction commitments if its bid is accepted. Thus, its selling decision should be constrained by the amount it can generate on its own by committing its units plus the amount of energy it would buy from the spot market.

$$PSell_k \le \left(W_{i,k} \cdot P_i^{Min} + PG_{i,k}\right) + PBuy_k \tag{26}$$

4.4.2.3 Other Constraints

Other UC constraints were discussed in Chapter-2, and those are still applicable to the genco in deregulated markets. Those are listed below.

- Ramp rate constraint on thermal units
- Minimum up and down-time constraints on thermal units
- Maximum and minimum generation limits
- Must run units are constrained to operate throughout the said period

In addition to these constraints, the genco may need to include some hydro related constraints in order to ensure an optimal schedule for its hydro resources.

4.4.3 Example

Consider a genco with 10 generating units and an installed capacity of 869 MW. Its capacity mix is as follows. 400 MW of nuclear, 225 MW coal-fired, 94 MW oil-fired and 150 MW of hydro units. The details are given in *Table 5*. The cost function of the generators is assumed to be of the form:

$$C_i(P_i) = CMin_i + GCst_i \cdot PG_i$$

Note that in the above cost function, PG_i denotes the power generated by unit *i* over and above its minimum generating limit. The stated cost function is linear and thus the generator's marginal cost is constant. Such a cost function is used for the sake of simplicity.

It is also assumed that the genco is equipped with a forecast of the spot market price for the next day. This will be denoted as the nominal price. Based on this nominal price, we construct two price scenarios, a low price scenario and a high price scenario to examine how generation scheduling and UC decisions change, with price. And finally, the genco also has available the bilateral contract prices which, as mentioned earlier, are usually fixed *a priori* and does not affect the model optimization (*Figure 11*).

Table 5. Unit specific data of the genco

Unit	Type	P^{Max}, MW	P^{Min}, MW	GCst, \$/MWh	CMin,\$	ST, \$	SD, \$	MUT, hrs	MDT, hrs
1	OFST	12	2.40	27.60	35.88	0	0	0	0
2	OFST	12	2.40	27.60	35.88	0	0	0	0
3	OFST	20	4.00	43.50	25.00	1	0	0	0
4	OFST	50	10.0	23.00	29.90	1	0	0	0
5	CFST	75	15.0	14.40	18.72	10	4	3	4
6	CFST	150	35.0	11.64	13.44	20	8	3	4
7	NUC	400	100.0	6.00	7.53	35	20	MR	MR
8	HYD	50	0.0	2.0	0.2	0	0	0	0
9	HYD	50	0.0	2.0	0.2	0	0	0	0
10	HYD	50	0.0	2.0	0.2	0	0	0	0

OFST: Oil fired steam turbine; CFST: Coal fired steam turbine
NUC: Nuclear; HYD: Hydro
MR: Must-Run Unit

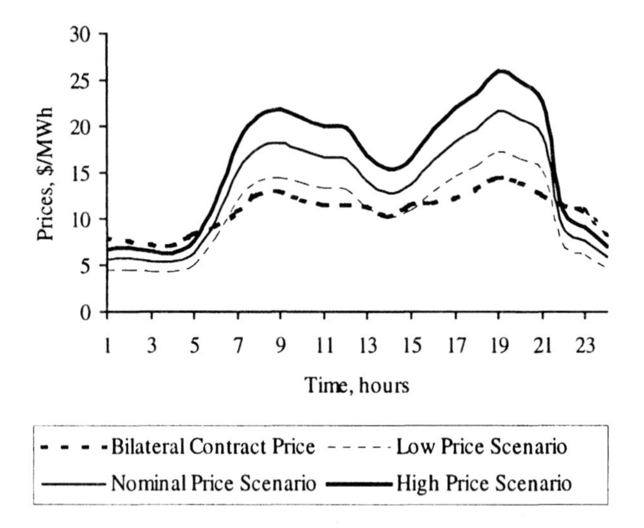

Figure 11. The bilateral contract price and the three price scenarios used in this example is shown. Note that the spot market price is assumed to be the 'nominal price scenario'. These figures are used to explain the example and do not resemble true market prices or their trends.

4.4.3.1 Comparison of the UC Schedule in Deregulated Environment with the Conventional Least Cost Schedule

The conventional UC discussed in Chapter-2 used an objective function that minimized the total system cost. The same is used here to compare how the genco's UC decisions change when spot-market participation is introduced. The cost function for a 24-hour scheduling period is given by,

$$C = \sum_{k=1}^{24}\left[\sum_{i=1}^{NG}\left[\begin{matrix}(W_{i,k}\cdot CMin_i + PG_{i,k}\cdot GCst_i) + UST_{i,k}\cdot ST_i \\ + USD_{i,k}\cdot SD_i\end{matrix}\right]\right] \tag{27}$$

The corresponding profit function of the genco over the 24-hour scheduling period, with cost minimization as an objective, may be written as:

$$\text{Profit} = \left[\sum_{k=1}^{24} CP_k \cdot BC_k\right] - C \tag{28}$$

Based on the unit commitment model for gencos operating in deregulated environment discussed in Sections-4.4.1 and 4.4.2, simulations are carried out considering the three scenarios of prices shown in *Figure 11.* A summary comparison of the operating and business decisions of the genco with the two different objective functions, given by (24) and (27), is shown in *Table 6*.

Table 6. Comparison of different objectives for a days operation

Operational Parameters	Minimum Cost (Conventional UC)	Maximum Profit (Modified UC)		
		Low Price Scenario	Nominal Price Scenario	High Price Scenario
Revenue earnings, $	145835.28	142191.24	160196.93	179257.655
Operation cost, $	66866.011	58781.92	75403.93	90138.545
Net Profit, $	78969.269	83409.32	84792.996	89119.11
Generation, MWh	14289.0	12820.29	14361.98	16068.14
Power sell in spot market, MWh	-	611.68	1285.91	1938.03
Power buy in spot market, MWh	-	2080.395	1212.93	158.89
Bilateral contracted demand, MWh	12990	12990	12990	12,990
Reserve energy, MWh	1299	1299	1299	1,299

We can see from *Table 6* that the genco increases its revenue earnings and profit by shifting over from a least-cost operation to a profit driven one. In the low price scenario, the genco chooses to buy from the spot market by de-committing some of its own generators; on the other hand, in the high price scenario it commits its generators so as to maximize profits through power sell in the spot market.

The unit commitment decisions in the cost minimization case (*Table 7*) show that the genco chooses to de-commit units 1 and 2 over a major part of the next day since they have fairly high operating costs, low start-up and shut down costs. On the other hand, in the competitive market environment, the genco's unit commitment decisions depend on the type of price scenario considered. The low price scenario produces similar commitment decisions as in the least-cost case since the genco now chooses to keep the costly generators off-line and purchase bulk amounts of energy from the spot market at low prices to maximize its profit.

In the nominal and high price scenarios, the unit commitment decisions change substantially, though. Units 1 and 2 are now committed over longer time periods, next day, particularly over the morning and evening peak hours.

There are also some differences in the scheduling of the hydro units across the scenarios. Though, in all cases, they are scheduled during the peak hours, the least cost schedule is more dispersed.

Table 7. Unit commitment decisions with cost minimizing objective (The elements in bold underline are those which change in the profit maximizing solution, nominal price scenario)

Hour / Unit	1	2	3	4	5	6	7	8	9	10	11	12	13	14	15	16	17	18	19	20	21	22	23	24
1	0	0	0	0	0	0	**0**	**0**	**0**	**0**	**0**	**0**	0	0	0	**0**	**0**	**0**	1	**0**	**0**	0	0	0
2	0	0	0	0	0	0	**0**	**0**	**0**	**0**	**0**	**0**	0	0	0	**0**	**0**	1	1	**0**	**0**	0	**1**	0
3	0	0	0	0	**0**	1	1	1	1	1	1	1	1	1	1	1	1	1	1	1	1	1	1	**1**
4	0	1	1	1	1	1	1	1	1	1	1	1	1	1	1	1	1	1	1	1	1	1	1	1
5	0	1	1	1	1	1	1	1	1	1	1	1	1	1	1	1	1	1	1	1	1	1	1	1
6	1	1	1	1	1	1	1	1	1	1	1	1	1	1	1	1	1	1	1	1	1	1	1	1
7	1	1	1	1	1	1	1	1	1	1	1	1	1	1	1	1	1	1	1	1	1	1	1	1
8	0	0	0	0	0	**1**	1	1	1	1	1	**0**	**1**	0	0	1	1	**0**	1	1	1	**1**	**1**	0
9	0	0	0	0	0	0	**0**	1	1	**0**	1	1	0	0	**1**	1	1	1	1	1	1	**1**	0	0
10	0	0	0	0	0	0	1	1	1	1	**0**	1	**1**	0	**1**	**0**	**0**	1	1	1	1	**1**	**1**	0

It is interesting to examine the total generation scheduled by the genco in the different price scenarios of the profit maximization case (*Figure 12*).

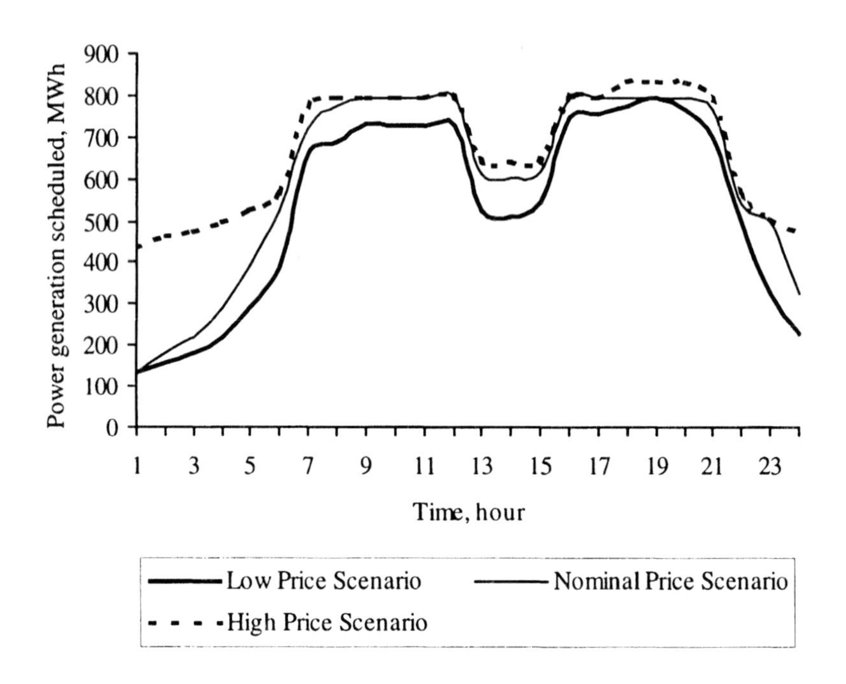

Figure 12. Total generation scheduled by the genco in different price scenarios of the profit maximization case

As we see in *Figure 12*, the high-price scenario schedules more generation than the low-price scenario during all hours. During hours 13-15, there is a slump in the scheduled generation across all scenarios. This is due to the role played by start-up costs, shut down costs and ramp-rate limits, which determine the units operation to a large extent. Here, it was cheaper for the genco to shut down some units during hours 13-15, than to generate and sell in the market. A scenario with an even higher price might have helped the genco cross the cut-off point in its revenue returns allowing it to start-up those shut-down units during hours 13-15.

On the other hand, a genco with high start-up and shutdown costs would prefer keeping its units running during the hours 13-15 and increasing its revenue by selling in the market.

4.5 Competitive Bidding

In a competitive electricity market, the sellers and buyers submit bids for energy buy and sell. In some markets, for instance in New Zealand, there is also a provision for simultaneous bidding for energy reserves [9]. However, in this section, we shall limit our discussion to energy bids only.

The bids are generally in the form of price and quantity quotations and specify how much the seller or buyer is willing to sell or buy, and at what price. After the bids are available to the market operator, the latter settles the market based on a criterion as discussed in the earlier sections of this chapter. Once the market is cleared, all selling participants receive a uniform price for their power delivered, *i.e.*, the market price, from the buying participants.

There are certain additional payments involved in some pools. For example, in UK there is a capacity payment component (*refer Section-2.1.1 for detail*). In the NordPool market too, the pool price is modified under certain circumstances, particularly when transmission congestion occurs. This will be discussed further in Chapter-4. However, here we restrict our discussions to the market price only.

This form of auction, where all winning bidders are offered the same price without any discrimination, and regardless of their individual bid, is known as *non-discriminated auction* or *second price auction*. This is usually the price of the highest priced bid that is cleared in the market. The non-discriminated auction provides incentives to bidders to bid their true costs and avoid guessing the bids of others.

On the other hand, in a *discriminated auction* or *first price auction* all bidders are not offered the same price after the market is settled. The bidders get the price that they had actually bid for, in the first place. A disadvantage of this system is that, it can give rise to *gaming* opportunities for the participants thereby providing ample scope for over-bidding and pushing up the market-clearing price.

4.5.1 Strategic Bidding

Construction of appropriate bids is very important for participants in the electricity markets since their underlying objective is to maximize profits. Strategic bidding is consistent within the principles of system operation and usually the participants do have the freedom to bid at prices different from their costs.

In [30] various power system requirements have been brought out which could be bid. The basic biddable components outlined are energy, capacity and reactive power. These have been further disaggregated into various categories, which come under the provision of ancillary services.

Bidding schemes for power exchanges have been developed in which buyers and sellers participate in auctions (even double or multiple auctions) by interacting through an Independent Contract Administrator [31 and 32]. In [31] a double auction game is used in combination with classical optimization and incorporating operational constraints to determine the energy transactions while in [32 and 33] a genetic algorithm is used to determine the bidding strategies.

4.5.1.1 Parameters Affecting Bidding Strategies

As stated earlier, most often a genco is free to design its bidding strategies based on its understanding of the market. However, in general, the following parameters are considered essential by a genco while formulating its bidding strategy:

- Technical constraints on unit operation

  These primarily include the UC constraints such as minimum up time and minimum down time, ramp rate limits, must run units, unit minimum and maximum generating capacity limits. An important consideration here is the type of market. As we have seen earlier, some markets require the gencos to bid their ramping prices also and unit commitment decisions are part of the market operator's functions.

  On the other hand, some markets require the gencos to carry out their own unit commitment, e.g. NordPool, and then the gencos' bids are formulated by internalizing the unit commitment decisions. The market operator does not have any role in those matters.

- Bilateral contracts

  The amount of bilateral contract scheduled for supply during the bidding period will determine how much power is available to the genco for bidding into the market. This holds true where the pool is not mandatory and the gencos are free to establish their own contracts with customers.

➢ Market clearing prices of the previous day or in the recent few days and also prices during that period of the year, during the past years.

This information, along with a good forecast of the next day's spot price, is a crucial input to the formulation of bidding strategies.

➢ Hydro energy availability for the next day

For those systems with a dominant share of hydro, this is an important factor. In Norway for example, the system is based on 100% hydro generation, and the system price in Norway is quite sensitive to the hydro energy availability (*Figure 13*).

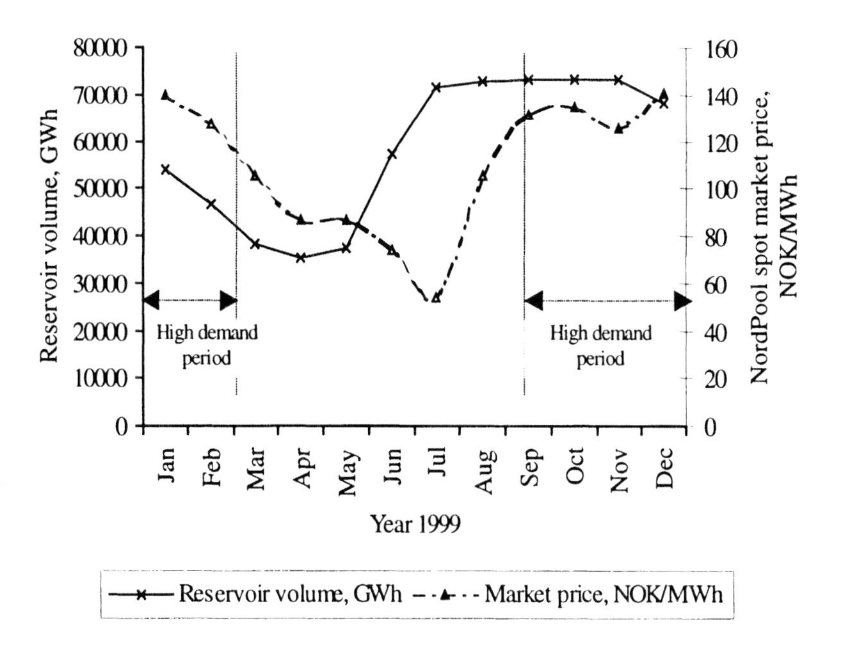

Figure 13. Average monthly hydro energy availability in the reservoirs in Norway vis-à-vis NordPool spot market prices at Oslo in 1999. *Note that the reservoir volume and market prices are on different axes.* (1 US$ ≅ 9 NOK as of February 2001) Source: www.nordpool.no

As we note from *Figure 13*, the reservoirs start filling up in Norway from the month of May and by August they are full. On the other hand this is the period of lowest demand on the system. The spot market prices went down drastically during the summer of 1999 (May to August) due to reduction in system demand coupled with good water availability. On the other hand, during the system peak demand in winter, the water availability is the highest and prices increase. Evidently there is a direct correlation between the system demand and market price. The important point to note here is that due to this correlation, during years of low rainfall, the impact on market prices can be severe. And it is for this reason that the market players in Norway use various estimates of water availability in the system especially during the winter months for strategic bidding.

- Load and weather forecast for the next day

- The genco should have a good idea of its unit generating costs and the system marginal cost characteristic as generation is increased.

4.5.1.2 Bidding Strategies

The problem confronted by a bidder in a power pool has been treated as an optimization problem that is solved to determine the optimal bidding strategies in [14, 16, 17, 34, 35, 36, 37] and by many others. The usual objective in this type of problem has been to consider the bidder's profit. In [34] a probability density function is used to characterize the competitor's bidding range and the optimal bid is the one that maximizes the lower bound on expected value of savings. In [14 and 35] an analytical formulation of the problem faced by a bidder in competitive power pool of England and Wales is reported. The optimal bidding strategy developed therein is for the case of perfect competition, taking into account uncertainty in load forecast and actions of competitors. One of the important findings of [14 and 35] is that, regardless of generation resources, costs and constraints, a genco's optimal profit-maximizing bidding strategy, is one which bids to supply at generation costs and maximum capacity.

4.5.1.2.1 Probability of Acceptance of a Bid

Now let us consider an example where a genco has to formulate its bidding strategy for the next day, given that, it is equipped with the following information while formulating its bidding strategy:

a. Forecasts of market price for all bidding intervals next day, say, μ_k
b. Marginal cost (MC) characteristic over its generating regime

To treat the problem in a simple manner, let us assume that the genco bids do not affect the system price in any way and therefore, the genco is a *price taker* in the market and has no market power. The bidding strategy to be adopted by the genco will depend on its MC for an hour *vis-à-vis* μ_k. If the genco requires a sell bid to have a high probability of acceptance, it should bid lower than μ_k. On the other hand, for a buy bid it should bid higher than μ_k.

If the genco requires selling *b* units of energy at an hour, how should it bid? As mentioned in [36], a good bidding strategy requires a compromise between bids with high potential profit returns but high risk of losing generating opportunities and bids with a high probability of winning in the auction but low or even negative revenue adequacy.

Let us assume that the actual spot market price, at an hour *k*, given by ρ_k is a normally distributed random variable with probability distribution function $f(\rho_k)$, mean at μ_k and standard deviation σ (29) (*Figure 14*).

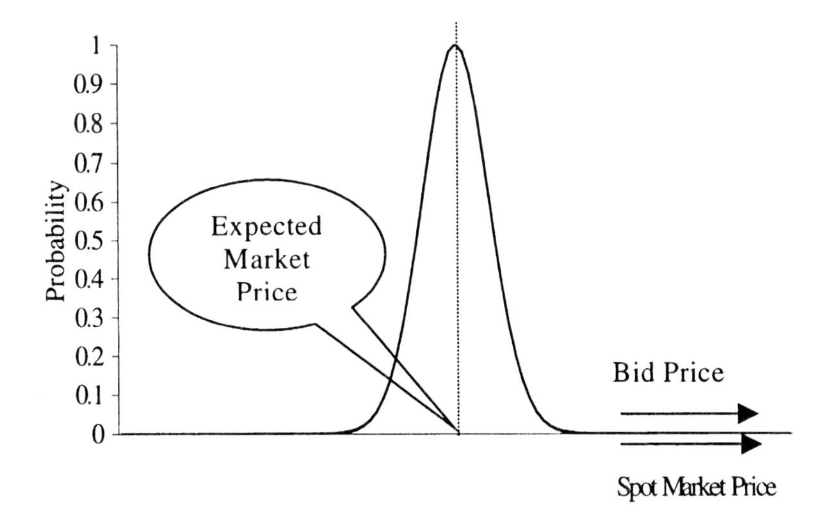

Figure 14. Probability of the location of actual market price vis-à-vis the forecast price

The probability distribution function can then be expressed as [38]:

$$f(\rho_k) = \frac{1}{2\sqrt{2\pi}} e^{-\left[\frac{(\rho_k - \mu_k)^2}{2\sigma^2}\right]}; \quad \text{where} \quad \int_{-\infty}^{\infty} f(\rho_k) d\rho_k = 1 \tag{29}$$

The cumulative distribution function of $f(\rho_k)$, $F(\rho_k)$ can then be given by:

$$F(\rho_k) = \int_{-\infty}^{BP_k} f(\rho_k) d\rho_k \quad = \frac{1}{2}\left[1 + \Phi(\alpha_{BP;k})\right] \tag{30}$$

$\Phi(\alpha)$ is the Laplace function or *probability integral* with properties $\Phi(0)=0$, $\Phi(-\alpha)= -\Phi(\alpha)$ and $\Phi(\infty)=1$ (31).

$$\Phi(\alpha_{BP,k}) = \frac{2}{\sqrt{\pi}} \int_0^{\alpha_{BP,k}} e^{-t^2} dt \quad \alpha_{BP,k} = \frac{BP_k - \mu_k}{\sigma\sqrt{2}} \tag{31}$$

$\Phi(\alpha)$ can be obtained from standard Laplace function tables [38].

$F(\rho_k)$ denotes the probability that ρ_k is less than or equal to bid price (BP) *i.e.*, $P(\rho_k \leq BP_k)$. Thus, the probability of a sell bid, with price BPS, being cleared in the spot-market when $\rho_k \geq BPS_k$ can be given by (32).

$$P_A(BPS_k) = P(BPS_k \leq \rho_k) = \frac{1}{2}\left[1 - \Phi(\alpha_{k,BPS})\right] \quad (32)$$

Similarly, the probability of a buy bid, with price BPB, being cleared in the spot market when $\rho_k \leq BPB_k$ can be given by (33) [29].

$$\begin{aligned} P_A(BPB_k) &= P(BPB_k \geq \rho_k) \\ &= 1 - P(\rho_k \leq BPB_k) = \frac{1}{2}\left[1 + \Phi(\alpha_{k,BPB})\right] \end{aligned} \quad (33)$$

This is explained in *Figure 15* where the two cases are plotted. The point 'O' denotes the expected system price for an hour, as available from a price forecast program. A sell bid with price 'E' (*on the upper plot*) will have a probability of acceptance given by the area 'EBCDE'. It is evident that as E increases, the probability of acceptance of the sell bid will reduce. On the other hand, consider the case of a buy bid (*lower plot*). In this case, a buy bid of bid price 'F' will have a probability of acceptance denoted by the area 'ABFA'.

It is apparent from (32) and (33) that to increase the probability of acceptance, the bidder should either bid below ρ_k to ensure a high probability of acceptance of a sell bid or bid above ρ_k to ensure a high probability of acceptance of a buy bid. However, there would be several other considerations while formulating the bid and one of the other important criteria would be the likely revenue / profit made from the bid.

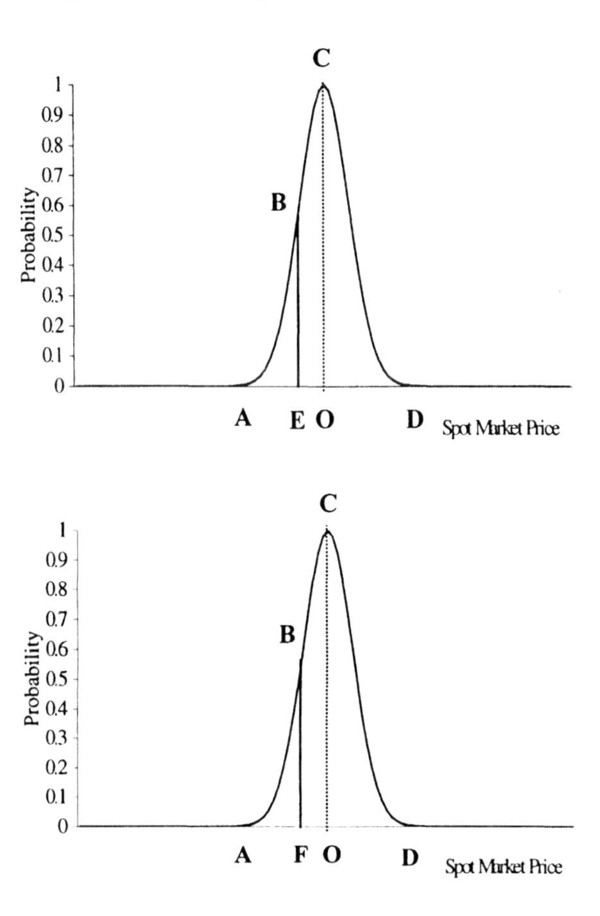

Figure 15. Determination of the probability of acceptance of selling (upper plot) and buying bids (lower plot). The plots assume a normal probability distribution of the actual market price about the expected market price (denoted by point O). Also note that, theoretically the points A and D are at infinity.

4.5.1.2.2 Other Considerations in Strategic Bidding

A strategic bidder requires considering several other factors while formulating its bidding strategy.

An interesting feature discussed in the context of the Spanish market is the possibility of generators to submit 'indivisible bids', *i.e.* bids which are accepted or rejected, but never partially dispatched. This ensures that the base load plants satisfy their must-run constraints as long as they bid low and the bids are accepted. There is also a provision to bid a 'minimum income' a genco is willing to receive. This ensures that the genco's costs are recovered as long as the bid is accepted.

It is interesting to note the UK experience that the base-load generators, particularly the nuclear units rarely bid their marginal costs. Most often, they bid very low prices and sometimes even zero prices, in order to get their bid cleared. This is done in order to ensure that their base-load units satisfy the 'must-run' constraint, discussed in the context of unit commitment in Chapter-2, and that they do not have to back down their units.

A contrary situation is found in the Australian market where the thermal (base-load) generators bid very high prices. This is because, these generators are mostly contracted to sell power through direct bilateral contracts and hence they do not have the concerns about 'must-run' status. Therefore, these gencos bid their excess power at very high prices to maximize their revenue. On the other hand since the gas-based gencos do not have bilateral contracts, they wait for the spot prices to increase [39].

It has been argued that the conventional practice of bidding based on marginal costs, with the assumption of a *perfect competition*, will not result in revenue adequacy for the bidder when used in competitive markets. In [36] a revenue adequacy constraint has been incorporated within the bidder's optimization framework.

An important aspect in developing bidding strategies is the consideration of *imperfect competition* in the market. This requires modeling competitor's behavior in response to their expectation of a genco's likely bids. In [40] coalition of bidders has been examined using a game theoretic approach and it is seen that when transmission constraints are binding, the benefits from coalition are high. In [41] a non-cooperative game theoretic approach is used to develop bidding strategies of gencos with incomplete information about the other participants. This helps the genco to develop its strategies considering its own perspective of the market as well as the perspectives that other participants have about it.

5. CONCLUDING REMARKS

Generation scheduling in deregulated electricity markets has undergone a change of philosophy and therefore it is required that we examine the new paradigms of operation. In this chapter we discussed the two basic players in such markets- the independent system operator (ISO) and the genco.

We have seen that many of the activities of the operator in conventional vertically integrated systems have now been shifted to the market operator or the genco. For example, the unit commitment and load dispatch responsibilities in bilateral markets are no longer the responsibility of the system operator. On the other hand some activities have been taken over by a market operator (though in some markets the ISO and the market operator remain the same entity).

Two basic market structures- the bilateral contract dominated market and the pool were discussed in the chapter. The role of the ISO and the genco in these market environments are different, and this has been highlighted. Settlement of markets using the objective of maximization of social welfare or minimization of consumer payment is discussed. Finally, we examine the various issues involved in formulation of bidding strategies.

REFERENCES

1. P. Varaiya and F. F. Wu, "MinISO: A minimal Independent System Operator", Proceedings of 30th Annual Hawaii International Conference on System Sciences, 1997.
2. H. Rudnick, R. Varela and W. Hogan, "Evaluation of alternatives for power system co-ordination and pooling in a competitive environment", IEEE Transactions on Power Systems, May'97, pp.605-613.
3. D. Shirmohammadi, B. Wollenberg, A. Vojdani, P. Sandrin, M. Pereira, F. Rahimi, T. Schneider and B. Stott, "Transmission dispatch and congestion management in the emerging energy market structures", IEEE Transactions on Power Systems, Nov.'98, pp.1466-1474.
4. R. D. Tabors, "Lessons from the U.K. and Norway", IEEE Spectrum, Aug.'96, pp.45-49.
5. R. D. Christie and I. Wangensteen, "The energy market in Norway and Sweden: Introduction", IEEE Power Engineering Review, Feb.'98, pp.44-45.
6. R. D. Christie and I. Wangensteen, "The energy market in Norway and Sweden: The spot and futures markets", IEEE Power Engineering Review, Mar'98, pp.55-56.
7. R. D. Christie and I. Wangensteen, "The energy market in Norway and Sweden: Congestion Management", IEEE Power Engineering Review, May'98, pp.61-62.
8. A. I. Cohen, V. Brandwajn and S. K. Chang, "Security constrained unit commitment for open markets", Proceedings of Power Industry Computer Applications Conference 1999 (PICA1999), pp. 39-44.
9. T. Alvey, D. Goodwin, X. Ma, D. Streiffert and D. Sun, "A security-constrained bid-clearing system for the New Zealand wholesale electricity market", IEEE Transactions on Power Systems, May '98, pp. 340-346.
10. X. Ma, D. Sun and K. Cheung, "Energy and reserve dispatch in a multi-zone electricity market", IEEE Transactions on Power Systems, Aug.'99, pp.913-919.
11. A. K. David, "Dispatch methodologies for open access transmission systems", IEEE Transactions on Power Systems, Feb.'98, pp.46-53.
12. K.W. Cheung, P. Shamsollahi and D. Sun, "Energy and ancillary services dispatch for the interim ISO New England electricity market", Proceedings of Power Industry Computer Applications Conference 1999 (PICA1999), pp. 47-53.
13. E. S. Huse, I. Wangensteen and H. H. Faanes, "Thermal power generation scheduling by simulated competition", IEEE Transactions on Power Systems, May '99,pp.472-477.
14. G. Gross and D. J. Finlay, "Optimal bidding strategies in competitive electricity markets", Proc. of 12th Power Systems Computation Conference, 1996, pp.815-823.
15. J. Perez-Arriaga, C. Meseguer, "Wholesale marginal prices in competitive generation markets", IEEE Transactions on Power Systems, May '97, pp. 710-717.

16. D. Zhang, Y. Wang and P. B. Luh, "Optimization based bidding strategies in the deregulated market", Proceedings of Power Industry Computer Applications Conference 1999 (PICA1999), pp. 63-69.
17. J. D. Weber and T. J. Overbye, "A two-level optimization problem for analysis of market bidding strategies", Proceedings of IEEE Power Engineering Society Summer Meeting 1999, Vol.2, pp.682-687.
18. S. Hao, G. A. Angelidis,H. Singh and A. D. Papalexopoulos, "Consumer payment minimization in power pool auctions", IEEE Transaction on Power Systems, Aug.'98, pp.986-991.
19. J. M. Jacobs, "Artificial power markets and unintended consequences", IEEE Transactions on Power Systems, May'97, pp.968-972.
20. J. Alonso, A. Trias, V. Gaitan and J. J. Alaba, "Thermal plant bids and market clearing in an electricity pool. Minimization of costs vs. minimization of consumer payments", IEEE Transactions on Power Systems, Nov.'99, pp. 1327-1334.
21. F. D. Galiana and M. Ilic´, "Framework and methods for the analysis of bilateral transactions", in M. Ilic´, F. Galiana and L. Fink edited: *Power Systems Restructuring: Engineering and Economics*, Kluwer Academic Publishers, 1998.
22. F. D. Galiana and M. Ilic´, "A mathematical framework for the analysis and management of power transactions under open access", IEEE Transactions on Power Systems, May'98, pp.681-687.
23. J. W. M. Cheng, F. D. Galiana and D. T. McGillis, "Studies of bilateral contracts with respect to steady-state security in a deregulated environment", IEEE Transactions on Power Systems, Aug.'98, pp.1020-1025.
24. J. Zhong and K. Bhattacharya, "Optimum Var support procurement for maintenance of contracted transactions", Proc. of International Conference on Electric Utility Deregulation and Restructuring and Power Technologies 2000, DRPT2000, London, 2000.
25. J. Bastian, J. Zhu, V. Banunarayanan and R. Mukerji, "Forecasting energy prices in a competitive market", IEEE Computer Applications in Power, July '99, pp.40-45.
26. O. B. Fosso, A. Gjelsvik, A. Haugstad, B. Mo and I. Wangensteen, "Generation scheduling in a deregulated system. The Norwegian case", IEEE Transactions on Power Systems, Feb.'99, pp.75-81.
27. B. R. Szkuta, L. A. Sanabria and T. S. Dillon, "Electricity price short-term forecasting using artificial neural networks", IEEE Transactions on Power Systems, Aug.'99, pp.851-857.
28. C. W. Richter and G. B. Sheblé, "A profit based unit commitment GA for the Competitive Environment", IEEE Transactions on Power Systems, May 2000, pp.715-721.
29. K. Bhattacharya, "Strategic bidding and generation scheduling in electricity spot-markets", Proc. of International Conference on Electric Utility Deregulation and Restructuring, and Power Technologies 2000, DRPT2000, London, 2000.
30. J. J. Ancona, "A bid solicitation and selection method for developing a competitive spot priced electricity market", IEEE Transactions on Power Systems, May'97, pp.743-748.
31. J. Kumar, G. B. Sheble, "Auction game in electric power markets", Proceedings of 58th American Power Conference, 1996, pp.356-364.
32. C. W. Richter and G. B. Sheble, "Genetic algorithm evolution of utility bidding strategies for the competitive marketplace", IEEE Transactions on Power Systems, Feb.'98, pp.256-261.

33. C.W. Richter, G. B. Sheble and D. Ashlock, "Comprehensive bidding strategies with genetic programming / finite state automata", IEEE Transactions on Power Systems, Nov.'99, pp.1207-1212.
34. J. W. Lamont, S. Rajan, "Strategic bidding in an energy brokerage", IEEE Transactions on Power Systems, Nov.'97, pp.1729-1733.
35. G. Gross, D. J. Finlay, G. Deltas, "Strategic Bidding in Electric Generation Supply markets", Proceedings of IEEE Power Engineering Society Winter Meeting 1999, pp.309-315.
36. C. Li, A. J. Svoboda, X. Guan and H. Singh, "Revenue adequate bidding strategies in competitive electricity markets", IEEE Transactions on Power Systems, May 1999, pp.492-497.
37. H. Song, C. C. Liu and J. Lawarree, "Decision making of an electricity suppliers bid in a spot market", Proceedings of IEEE Power Engineering Society Summer Meeting 1999, Vol.2, pp.692-696.
38. E. Suhir, *Applied probability for engineers and scientists,* McGraw Hill, 1997.
39. W. Mielczarski, G. Michalik and M. Widjaja, "Bidding strategies in electricity markets", Proc. of Power Industry Computer Applications Conference 1999 (PICA1999), pp.39-44.
40. R. W. Ferrero, S. M. Shahidehpour and V. C. Ramesh, "Transaction analysis in deregulated power systems using game theory", IEEE Transactions on Power Systems, Aug.'97, pp.1340-1347.
41. R. W. Ferrero, J. F. Rivera and S. M. Shahidehpour, "Application of games with incomplete information for pricing electricity in deregulated power pools", IEEE Transactions on Power Systems, Feb.'98, pp.184-189.

Chapter 4

Transmission Open Access and Pricing Issues

1 INTRODUCTION

The competitive market for electricity has developed at two ends, the generation end and the retail supply end, while the transmission sector remains a monopoly business and therefore regulated. The transco facilitates the trading between parties and has therefore played a vital role in the restructuring of the power industry in all countries. As we also know, power transmission is an area where economics of scale are quite significant and hence there are tendencies in the system to produce monopolies.

The basic parameters that characterize the electric power transmission sector are [1]:

- large sunk and lumped investments
- need for redundancies to meet security requirements
- economics of scale in the construction cost in terms of the capacity of the transmission line
- economics of scope given by the interconnection of electric systems

In order to prevent the transco from overcharging for the service, there is a need for the transmission systems to be regulated. The need for regulation is all the more important when the transmission grid is the nucleus of competition among geographically dispersed generators. The trend of establishing new legal and regulatory frameworks offering third parties open access to the transmission network may be seen as a logical outcome in this context.

1.1 The US and the European Perspective: Transco vis-à-vis the Transmission System Operator

It is important to briefly explain the difference between the two terms that have figured in the discussions on transmission open access, one in the context of deregulation of the US electricity industry and the other in the context of the European deregulation. In US, the utilities have traditionally been the, so-called, vertically integrated systems with generation, transmission and distribution being under the aegis of one entity. With the passing of the (US) Power Utilities Regulatory Policy Act (PURPA) in 1978 and the emergence of non-utility generators (NUG) and gencos that were allowed to sell their excess power to the vertically integrated utilities, the access to transmission was opened up.

Subsequently with deregulation, the transmission activities were separated from those of generation and three new players- the gencos, the transcos and the discos came into being, along with a central coordinator the independent system operator (ISO). We have already discussed the role and functions of the ISO in the different market structures in Chapter-3.

On the other hand, the process of deregulation in Europe has been somewhat different from that in the US. Most of the cases of restructuring in Europe came about through the break-up of a large federal government-owned utility leading to the creation of a transco and a genco. The transco also assumed the role of the ISO, while at the same time owning the high voltage transmission network. The genco, on the other hand, operated as a competitive player along side other gencos in the market.

The split of the Central Electricity Generating Board in UK to form the National Grid Company, that of Vattenfall in Sweden to create Svenska Kraftnät or that of Statkraft in Norway to form Statnätt are some examples where the transmission system operator emerged from the state owned utility.

Therefore, as we see, the European structure has been moving towards the model of a unifying entity, that simultaneously plays the roles of a transco and an ISO. Such an entity has been termed as the Transmission System Operator (TSO) in the European context. In the UK system, in addition to its above mentioned roles, the TSO has the additional responsibility of carrying out the market settlement through a centralized dispatch based on a bidding process, discussed in Chapter-3.

2 WHAT IS POWER WHEELING?

Exchange of power between utilities has been practiced even before the term *wheeling* came into existence. However, with the passage of the PURPA in 1978, there was a tremendous increase in NUG generation capacity. It is estimated that during the eighties decade, the NUG capacity addition in US was approximately 30,000 MW and was expected to account for between 20 to 40 percent of all additional new capacity during 1991-2000 [2]. A direct consequence of this increase in capacity was an increase in power transactions involving NUG. Understandably, most of these transactions involving NUG, involve customers who are not connected to the former directly.

This led to a significant increase in power wheeling transactions, defined as the transmission of active and reactive power from a seller to a buyer through the transmission network owned by a third party. The third party is the wheeling utility, which carries this power and should be paid for this service since it has to bear the costs of losses from such transactions. Also, such wheeling of power may require the wheeling party to reschedule its own generation thereby increasing its costs.

Therefore, the important issues pertaining to power wheeling relate to the sharing of benefits from these transactions and the pricing of the power transactions.

With the increase in the amount of energy transactions, the evaluation of transmission capacity used for the same is an important problem for the system operator. In [3], a MW-mile based methodology has been described which evaluates the actual usage of the transmission network.

Wheeling transactions have been categorized in [4] into four types depending on the location of the seller and the buyer. The first, *bulk power wheeling* denotes power transactions between two fully regulated utilities using the network of a third utility. *Customer wheeling* is when an independent customer purchases power from a utility through the intervening network of another party while *supplier wheeling* is when an independent generator sells to a utility through a third party network. The fourth type is found when an independent generator sells power to an independent customer through the wheeling network, called *supplier-to-customer wheeling*.

3 TRANSMISSION OPEN ACCESS

Wheeling is thought of as a one-time isolated service of delivering power between two parties by a third party. On the other hand, transmission open access and provision of related services is a separate business in itself, providing and facilitating electricity market competition and hence, requires to be treated separately.

According to the IEEE Task Force on Transmission Access [5], the term 'transmission access' refers to the requirement that the transmission network owners make their systems available to the other players in the system. That may include independent generators, customers, or other utilities, that may desire to use the network for power transactions between themselves. A set of comprehensive surveys carried out by the IEEE Task Force provides the large volume of work available in the literature on issues pertaining to transmission open access- operational, planning, reliability, costing, pricing and regulatory [6, 7].

The background to this was the (US) Energy Policy Act of 1992 that laid the foundations for the provision of open access, and that subsequently proved to be a crucial catalyst in the creation of a competitive electricity market.

The implications of transmission open access could be varied and may span over a wide range right from those pertaining to ancillary services, such as participation of independent generators as ancillary service providers, up to the opening up of the distribution networks to competition, the so called retail competition. It also impacts the operations planning activities of the ISO, more so in case of the pool model in which all generation scheduling is centralized.

3.1 Types of Transmission Services in Open Access

As noted in the US Energy Policy Act of 1992, the transmission utility should recover all such costs incurred in connections and in providing for transmission and other necessary associated services. This is a somewhat unresolved issue, with the unbundling of various services with deregulation. The emergence of provisions for ancillary services and the various classifications and definitions that go with it, tend to make the cost recovery issue quite unclear.

In a competitive market environment, there are various types of transmission transactions and the cost of a transaction will depend on the type of transaction carried out and the cost components considered. Based on an understanding of the costs, pricing for these services can be subsequently analyzed.

The basic categories into which transmission services can be classified are (a) point-to-point services and (b) network services. While the point-to-point services are those with specified delivery and receipt points, the network services allow the transmission user a complete access to the system with no specification on the points of delivery or receipt, nor any additional charge for change of schedules.

The point-to-point services can be further classified as shown in *Figure 1* and each of the transactions can be explained as follows [8, 9].

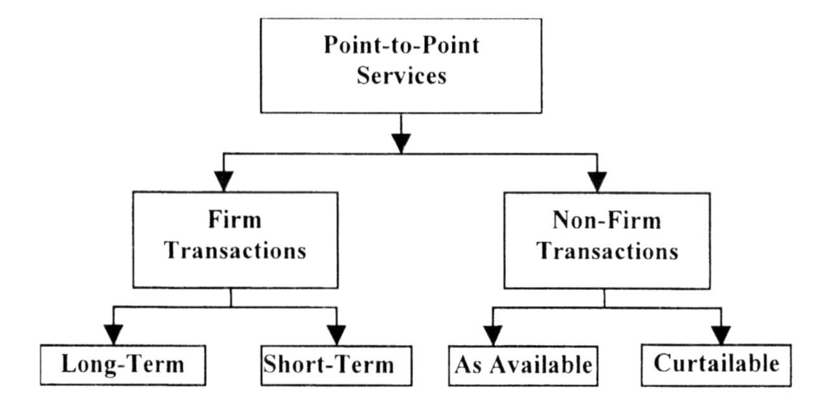

Figure 1. Classification of Point-to-Point Transmission Transactions

Firm transactions- these transactions are not subject to discretionary interruptions and are specified in terms of MW of transmission capacity that must be reserved for the transaction. The transco makes arrangements for enough capacity on the network to meet these transaction needs.

These could either be on a long-term basis, in the order of years, in which case the charges for such transactions can be designed to incorporate capacity investment needs of the network, or on short-term contracts (up to one year).

Non-firm transactions- these transactions are not firm and hence are subject to curtailment from the transco in times of network congestion, outages and overloads, or even on the basis of economic opportunities to the transmission provider.

The as-available transactions are basically isolated contracts on a short-term basis as per availability of transmission capacity while curtailable transactions may be regular transactions with a different pricing policy since these would not be charged for the capacity investment components.

4 COST COMPONENTS IN TRANSMISSION

In this section we briefly discuss those costs that a transco incurs in order to fulfill the transmission contracts satisfactorily. A comprehensive analysis of the costs incurred and methods on how to determine such costs have been provided in [9].

- Operating Cost

  These costs are incurred by a transco in carrying out the transactions generally relating to the cost of rescheduling of generation as well as those related to maintaining the system voltages, reactive power support and line flow limits.

  In those systems where the transco is also the central authority in dispatching generation, the costs can be obtained by appropriate OPF simulation where the operating cost is the dual of the demand-supply constraint. This would require a series of two OPF simulations, one without the transactions and one with the transactions and the cost difference incurred is then calculated. Another approach is to use the bus marginal costs.

- Opportunity Cost

  These costs are associated with the benefits, which the transco has to forgo in order to provide a transmission service. The unrealized benefits can be because the transco could not use cheaper generation resources due to transmission overloads. Or the unrealized revenue from those firm transactions, that could not be supported due to the operating constraints being binding.

- Reinforcement cost

  This is the capital cost of new transmission facilities needed to accommodate a transmission transaction. This component applies only to firm transactions.

- Existing Cost

  This is the cost of existing facilities associated with the investment already made in the system. This cost needs to be allocated to the transmission transactions on a rational basis.

Generally the total cost of a transmission transaction is the sum of the above four components, though the component that actually figures in the cost is determined from the type of transaction.

5 PRICING OF POWER TRANSACTIONS

The objective of any transmission-pricing scheme is to allocate all or part of the existing and new cost of transmission system to the customers. However, tariffs for transmission services are more often set by government regulations, and are based on its policy directives. In spite of this, any transmission pricing strategy should seek to achieve the following basic goals:

Recover costs: The tariff charged for use of transmission services must produce enough revenue to cover all the expenses made in investment, operation and maintenance of the transmission network, as well as provide a small (regulated) level of profit for the owners.

Encourage efficient use: The price structure should give incentives for using the transmission system efficiently. Efficient use could mean ensuring both, economic efficiency by maximizing social benefits and technical efficiency by minimizing losses.

Encourage efficient investment: The price structure and the way money is paid to the owners should provide an incentive for investment in new facilities, when and where they are needed.

Fair: Must be fair and equitable to all users.

Understandable: All users must be able to understand the pricing structure.

Workable: The pricing scheme should be implementable in the actual system.

The pricing of power transactions has always been an important topic and the theory of marginal cost was exploited to develop pricing mechanisms for wheeling transactions. It is generally accepted that the transmission utility should be paid for use of its network based on the short-run cost of wheeling. This can be computed from the difference in marginal costs of electricity at the buses where it enters and leaves the transmission utility. Therefore an ideal wheeling rate was defined as [10],

$$\text{Ideal wheeling rate} = \text{Marginal cost of wheeling} \tag{1}$$

The wheeling rate in (1) recognizes transmission constraints as well as incremental loss components because of the detailed network representation with the inclusion of load flow equations in the model framework.

The above means that, if the power flow on a line is at the transmission limit, the wheeling rate is appropriately modified to include the dual of the transmission constraint in the marginal cost. Similarly, if a wheeling transaction results in reduction of system losses vis-à-vis system costs, the marginal cost of wheeling and the wheeling rate is negative.

In [11] a computer program, WRATES: *Wheeling Rate Evaluation Simulator* that uses a dc load flow representation for the network to determine the marginal cost of power wheeling has been discussed. Different wheeling rate contracts, including ones with different update frequencies, such as, one-hour update wheeling contracts, twenty-four hour update and time-of-use wheeling rates can be handled.

The transmission losses in wheeling transactions can have a significant role and in certain circumstances can even negate the total earning of the wheeling utility from the transactions, if they are not appropriately accounted for in the wheeling rates. In [12], on-line loss associated with a wheeling transaction is calculated and its effect on the production cost has been examined.

Another important issue in power transactions is the reactive power flows associated with them. These flows are important since they directly affect the third party's transmission losses and voltage levels. As identified in [13 and 14], though the marginal cost of reactive power generation is smaller compared to real power generation, the cost difference between entry and exit buses for both are comparable. Therefore, it is important to formulate separate wheeling rates for such reactive power flows.

For evaluation of reactive power wheeling costs, an OPF type framework is needed which can include the reactive power balance and handle the reactive constraints as well as voltage limits at bus bars. The marginal cost for real and reactive power can then be simultaneously determined, as described in Chapter-2, using the Lagrange multipliers of the corresponding load-flow equations.

The wheeling rate is therefore the difference between marginal cost of delivering (real or reactive) power at the buying bus and the selling bus [13].

$$\omega_P = \rho_{P_B} - \rho_{P_S}; \quad \omega_Q = \rho_{Q_B} - \rho_{Q_S} \tag{2}$$

ω_P and ω_Q are wheeling prices for real and reactive power respectively. ρ_P and ρ_Q are marginal costs of real and reactive power, as discussed in Chapter-2, while subscripts B and S represent the buyer bus and the seller bus respectively.

In Section-4, we examined the various cost components involved, while a transco provides its transmission facilities to the different market players. An appropriate pricing scheme would therefore ideally be one that recovers these prices from the users of the system in a rational manner.

To this effect, pricing schemes for transmission have evolved around three basic philosophies (a) the embedded-cost based paradigms (b) those based on incremental costs and (c) schemes which use a combination of (a) and (b).

These pricing paradigms have been brought out succinctly in [15]. The embedded cost based pricing schemes are based on the total transmission cost allocation to various transactions while the incremental cost based pricing seeks to identify the additional burden on a transmission system from one particular transaction.

It has been argued that none of these paradigms recover enough revenue to cater for the transmission costs already sunk, *i.e*, the investment costs in infrastructure. Therefore, the third paradigm combines the two earlier notions to evolve pricing schemes that recover both, the costs embedded in the system and those incurred by the system from one additional transaction. The following discussions on pricing paradigms are based on the unifying concepts developed in [15].

5.1 Embedded Cost Based Transmission Pricing

These methods consider the embedded capital costs and average annual operation costs of existing facilities while determining the transmission costs. For each transmission line, the net plant cost is calculated for each year of the transaction period. This is calculated using the replacement cost, average service life and depreciation reserve of the line capital investment. Subsequently, the annual fixed charge rate is calculated for each year [2, 16].

Based on these calculations, four different cost allocation methods are discussed in [2] namely, the rolled-in embedded cost (or postage stamp method), contract path, boundary flow and line-by-line (or MW-Mile) method. The first two do not require any power flow simulation, and are thus simple to handle. However for the same reason, the tariffs, which evolve in these two methods might be completely different from power flows actually taking place.

One shortcoming of the embedded cost based methods is that, these do not recover the cost of new transmission facilities or reinforcements and are hence not economically efficient. Two of the broadly used schemes within this paradigm, *viz.*, the *postage stamp method* and the *MW-mile method*, are discussed below.

5.1.1 Postage Stamp Method

The name of this scheme has understandably, evolved from the basis on which postage stamps are priced, *i.e.* the customer only pays according to the weight of the package, not on the basis of distance of delivery point, or how the package will contribute to the postal transport requirement, *etc.*

Evidently, this is the simplest method of transmission pricing and no distinction is made between transactions with regard to the power flow path, supply or delivery points, or the time when it takes place. Therefore, a transaction of 10MW between two adjacent buses could end up paying more than one between far off locations if the peak load on the system, in the later case is more. It also does not take into account the increased transmission losses due to a transaction. This method is simple to handle, though not very sound economically. Postage stamp rates are charged at a flat rate on per MW basis. Mathematically the transaction price in this scheme can be written as given in [15]:

$$R_t = TC \cdot \frac{P_t}{P_{Peak}} \tag{3}$$

R_t transmission price for transaction t in $

TC total transmission charges in $

P_t transaction t load at the time of system peak load condition in MW

P_{Peak} is system peak load in MW

5.1.2 MW-Mile Method

Two shortcomings of the postage stamp based pricing scheme, namely, not accounting for flows on specific lines, and not accounting for the distance of the transaction (and implicitly the losses), are addressed while devising the MW-Mile method.

In this method, the basic concept is that the power flow-mile on each transmission line due to a transaction is calculated by multiplying the power flow and distance of the line. The total transmission system use is then the sum of all the power flow-miles and this provides a measure of how much each transaction uses the grid. The price is then proportionate to the transmission usage by a transaction. This may be expressed mathematically as follows [15, 17]:

$$R_{T_i} = \sum_j \left[\frac{P_{j;T_i} \cdot L_j \cdot F_j}{\sum_i P_{j;T_i}} \right] \tag{4}$$

R_{T_i} is the price charged for transaction T_i in $/MW

$P_{j;T_i}$ is the loading of line j due to transaction T_i, MW

L_j is the length of the line j, mile

F_j is a pre - determined unit cost reflecting the cost per unit capacity of the line, $/MW - mile

Apparently, this method seeks to allocate the costs based on actual system use as closely as possible though there have been arguments that it suffers from defects arising from the lumping of operating and embedded costs.

A transmission-pricing scheme has been developed in [17] that accounts for the cost recovery of both embedded costs and reliability benefits from transmission system.

5.1.2.1 Example

Consider an eight-bus system comprising three generators at buses 1, 2 and 3 and three load buses namely 4, 7 and 8 (*Figure 2*). There are two transactions taking place in addition to the 'native' loads, which the transco has to satisfy:

a. A genco at bus 5 has entered into a contract with a customer at bus 6, for power sell of 300 MW (say T1).
b. A genco at bus 8 has entered into a contract with a customer at bus 4, for power sell of 75 MW (say T2).

The power will be transported through the network system and the transco will receive payments for power transactions from the buying / selling parties.

The total system demand is 450 MW ordinarily. The transmission system is owned and operated by a transco that is also responsible for the generation dispatch functions with a cost-minimizing objective.

Determine the transmission charges that the transco will receive, using the MW-Mile method. The transmission network parameters are taken from [13] and given in *Table 1* and the cost data for the generators are given in *Table 2*.

Table 1. System network parameters

Line i-j	Resistance, ohms	Reactance, ohms	Charging, p.u.
5-4	0.065	0.05	0.0
7-6	0.028	0.030	0.06
3-5	0.008	0.031	0.06
2-8	0.008	0.037	0.06
1-6	0.008	0.037	0.06
3-6	0.028	0.064	0.06
1-4	0.007	0.015	0.06
5-8	0.008	0.037	0.06

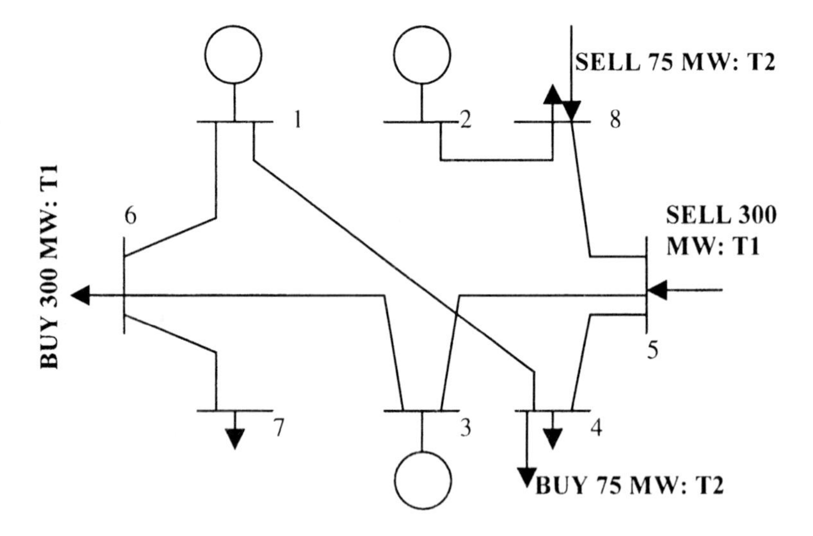

Figure 2. Example system for transmission pricing problem

The cost function for the generators is represented by a quadratic function of the generation as follows:

$$C_i = a_{o_i} P_i^2 + b_{o_i} P_i + c_{o_i}$$

Cost coefficients a_o, b_o and c_o for the three generators are given in *Table 2*.

Table 2. Generating unit cost coefficients and capacity limits

Generator	a_o	b_o	c_o	P^{Max}	P^{Min}	Q^{Max}	Q^{Min}
1	0.20	2.25	0.500	500	100	375	-100
2	0.25	2.81	0.444	200	40	200	-50
3	0.30	3.19	0.406	200	40	200	-50

In addition let us assume the line lengths in order to evaluate the transmission prices. Also, let us assume a unit cost ($c_{i,j}$) of 100$/MW-mile as the cost per MW per unit length of the line[1].

Table 3. Length of the transmission lines, $L_{i,j}$, in miles

Line, *i-j*	5-4	6-7	3-5	2-8	1-6	3-6	1-4	5-8
Length	500	200	310	470	300	640	150	370

The base case power flows (without any buy-sell transaction), the power flow for a 300 MW power transaction case (transaction T1), and the power flow for a 75 MW power transaction case (transaction T2) is shown in *Figure 3*, *Figure 4*, and *Figure 5* respectively.

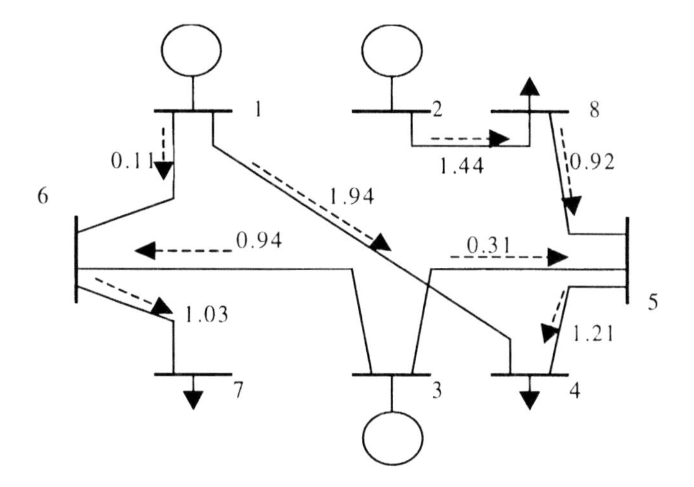

Figure 3. Line power flows in base case with a system demand of 450 MW. All power flows are in p.u. on a 100 MVA base. The loads are: 50 MW at bus-8, 300 MW at bus-4, 100 MW at bus-7.

Using the unit cost of the transmission lines of 100$/MW-mile and the transmission line length provided in *Table 3*, the solution is worked out as per the steps given below and corresponding calculations in *Table 4*. These calculations follow the method described in [3 and 15] except that the power flows are calculated using an OPF instead of a dc power flow. A good description of the method is provided in [3].

[1] Note that these values are only for the purpose of demonstrating the pricing method and may not relate to actual system costs.

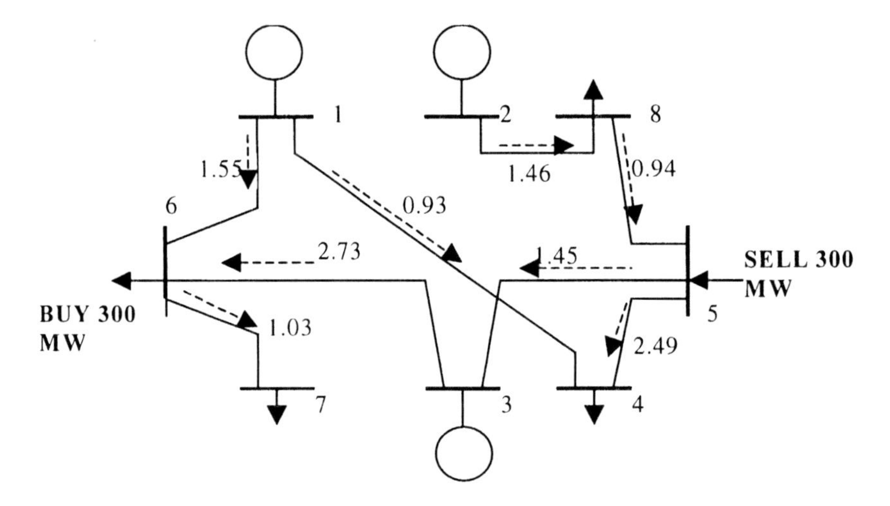

Figure 4. Line power flows due to a power transaction of 300 MW between buses 5 and 6

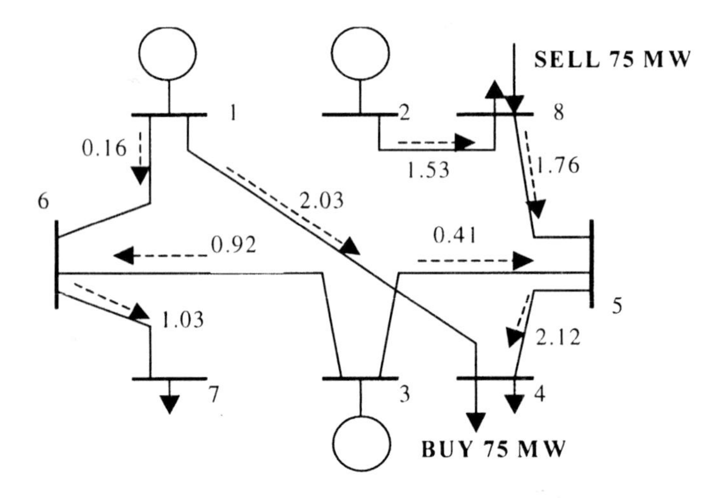

Figure 5. Line power flows due to a power transaction of 75 MW between buses 8 and 4.

Step-A: Find the total cost of the line by multiplying the unit cost of the line (assumed 100$/MW-mile for all lines) with the line length given in *Table 3.*

Step-B: Find the base case power flow on all lines. This is obtained here using an OPF as described in Chapter-2.

Step-C: Find the new power flow solution with the transaction T1, and hence the power flows on each line. (A negative sign indicates a reversal of flow direction vis-à-vis the base case flow).

Step-D: Find the new power flow solution with the transaction T2, and hence the power flows on each line.

Step-E: Calculate the incremental power flow on each line caused by the transaction T1.

Step-F: Calculate the incremental power flow on each line caused by the transaction T2.

Step-G: Calculate each line usage due to transaction T1 and hence find the total transmission system usage by T1.

Step-H: Calculate each line usage due to transaction T2 and hence find total transmission system usage by T2.

Step-I: Calculate the total transmission system usage by T1 and T2 for proportional allocation of the costs.

Step-J: Calculate the proportional allocation of cost to transaction T1.

Step-K: Calculate the proportional allocation of costs to transaction T2.

Table 4. Calculation of cost allocation based on the MW-Mile method. The power flows are from the ith bus to the jth bus. Negative sign on a P_{ij} power flow, denote reverse direction flow.

Steps	Line, i-j	5-4	6-7	3-5	2-8	1-6	3-6	1-4	8-5	Total
A	$L_{i,j} * c_{i,j}$	50,000	20000	31000	47000	30000	64000	15000	37000	
B	$P_{i,j}$ Base	1.21	1.03	0.31	1.44	0.11	0.94	1.94	0.92	
C	$P_{i,j}$ due to T1	2.49	1.03	-1.45	1.46	1.55	2.73	0.93	0.94	
D	$P_{i,j}$ due to T2	2.12	1.03	0.41	1.53	0.16	0.92	2.03	1.76	
E	F_j due to T1	1.28	0	1.14	0.02	1.44	1.79	1.01	0.02	
F	F_j due to T2	0.91	0	0.10	0.09	0.05	0.02	0.09	0.84	
G	A * E	64000	0	35340	940	43200	114560	15150	740	273930
H	A * F	45500	0	3100	4230	1500	1280	1350	31080	88040
I	$G_{Total} + H_{Total}$					361970				
J	G_{Total}/I					0.757				
K	H_{Total}/I					0.24.3				

As per the above calculations, the transaction T1 should pay 75.7% of the total costs while the transaction T2 for the remaining 24.3%.

An important point that needs mention here is that, the line usage from an additional transaction can be determined from various references points. For example, three methods to find the difference power flow due to a power transaction is discussed in [18], which are briefly mentioned here:

(a) The net difference method: |TF| - |BF|
(b) Vector difference method: |TF - BF|
(c) Positive difference method: ||TF|-|BF|| for TF > BF

TF is the transfer flow due to the power transaction and BF is the base power flow. In the present example, the vector difference concept was used to calculate the difference power flow due to a transaction.

5.2 Incremental Cost Based Transmission Pricing

5.2.1 Short-Run Marginal Cost (SRMC) Based Pricing

The short run marginal cost of a transco is the cost incurred in supplying an additional 1 MW of power in a transaction. This can be calculated from the difference in marginal costs at the supply bus and the delivery bus. This requires a complete network representation of the system in lines of an OPF and the dual variables associated with the demand balance equation (load flow equation) denote the marginal costs [3, 13, 17, 19]. The transaction price can be determined by multiplying the power transaction with the marginal cost to evolve the SRMC based price.

5.2.1.1 Example

Consider the eight-bus system described in the example in Section-5.1.2.1 and shown in *Figure 2*. Assume that the transmission system owner is also responsible for the dispatch of the generating units. Let us consider only transaction T1 now. In order to fulfil the transaction requirements, the transmission operator has to carry out re-dispatch of its generators, which involves additional costs.

Determine the transmission prices for the transaction based on the principle of short-term marginal costs.

Solution

The problem requires handling of both real and reactive power flows and constraints, and is solved using an optimal power flow model. We consider various cases with increasing power transaction on T1, starting from a no transaction case (*Table 5*).

The transmission owner schedules the generating units according to the least cost criterion. The marginal cost of real and reactive power at the selling bus (#5) and the buying bus (#6) are shown for different values of transaction steps. It can be seen that the system losses increase as transaction power increases. The system loss is met by the transmission owner by rescheduling its generators, thereby increasing the system costs.

Table 5. Optimal generation schedules with transaction T1 between bus 5 and 6

Case	Transaction T1, MW	P_1, MW	P_2, MW	P_3, MW	System Cost, $	System Loss, MW	MCP_B - MCP_S, $/MWh	MCQ_B - MCQ_S, $/MVAr
1	0	205	143	124	19339.2	21.3	6.68	0
2	25	207	142	124	19520.6	23.4	7.82	0
3	50	210	142	124	19730.6	25.9	8.98	0.040
4	75	212	142	124	19972.9	28.7	10.18	0.267
5	100	215	142	125	20250.2	31.9	11.42	0.514
6	125	218	142	125	20564.1	35.5	12.72	0.783
7	150	222	142	126	20916.6	39.4	14.09	1.079
8	175	225	142	126	21310.0	43.8	15.53	1.404
9	200	229	142	127	21746.9	48.7	17.06	1.761
10	225	234	143	128	22230.5	54.0	18.70	2.157
11	250	238	143	129	22764.2	59.7	20.46	2.595
12	275	243	143	130	23352.0	65.9	22.34	3.084
13	300	249	144	131	23998.6	72.6	24.38	3.629

MCP: Marginal cost of real power; MCQ: Marginal cost of reactive power
Subscripts B and S denotes the Buying and Selling buses respectively.

The power flow on each line of the transmission system for the case with T1 = 300 MW was shown earlier in *Figure 4*, while *Figure 3* showed the power flow on lines in the base case, *i.e.*, no transaction case. *Figure 6* shows the variation in system cost and loss as transaction power over the transmission system is increased.

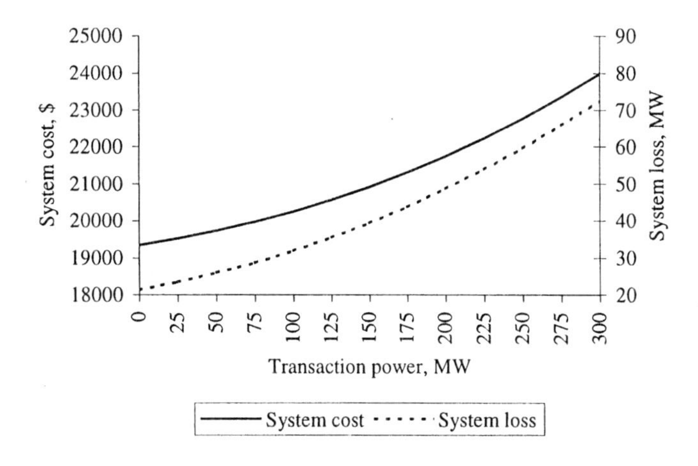

Figure 6. Variation of system cost and loss with power transaction T1

Based on the difference in the marginal costs of supplying real and reactive power between the buying and selling buses, the respective transmission prices for real and reactive power can be determined, as given by (1). This rate when multiplied by the transaction power gives the total price payable to the transco. *Figure 7* shows the total transmission price payable for different magnitudes of power transactions.

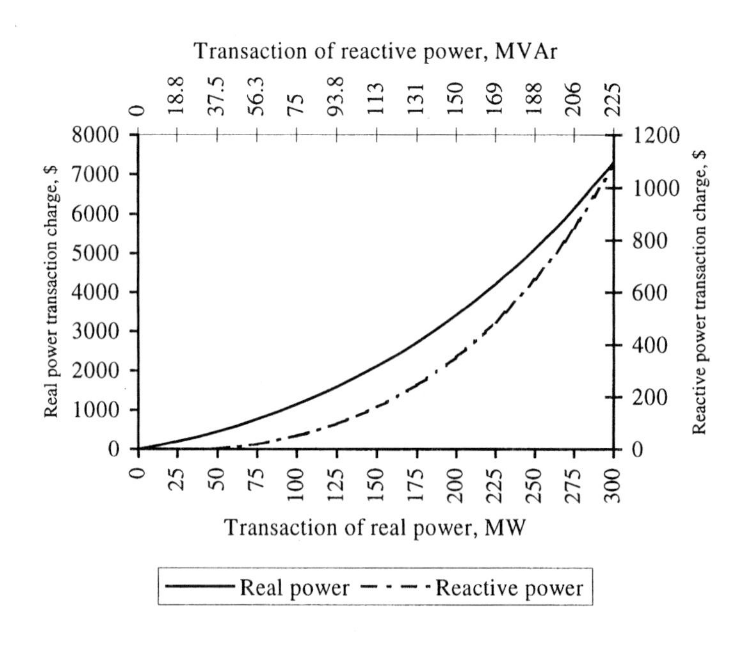

Figure 7. Total wheeling charge payable to wheeling utility-A

5.2.2 Long Run Marginal Cost (LRMC) Based Methods

In these methods, a long-term transmission planning analysis is included within the transactions. The models simultaneously determine the new transmission line additions over a period of time (about few years) and how and which power transaction brought about these additions. The Lagrange multipliers from such a model will provide the *long-run marginal cost* which is used for pricing to address the issue of capital cost recovery.

However, it should be noted that the computational complexity of such models is much higher due to the presence of integer variables (in new line selection), and the dynamic planning framework (due to the multiple years planning consideration).

In case of multiple power transactions taking place, this long-term marginal cost needs to be allocated fairly to all the transactions. In [2], four methods of long run marginal cost allocations have been suggested namely, dollar per MW, dollar per MW-mile, interface flow allocation by regions and one-by-one allocations.

5.2.2.1 Revenue Reconciliation

It is well known that short run marginal cost based pricing does not provide the revenue reconciliation to the transco and does not provide for cost recovery of capital investments made in transmission or capital costs of providing for transmission reinforcements. A detailed analysis of this issue [20] ascertains the reasons for this. It is shown that this occurs due to the discrepancies between planning horizons (*i.e.*, static vis-à-vis dynamic plans), plan deviations, strongly discrete nature of investments, economics of scale, reliability constraints and other constraints on network investments.

In the context of wheeling rates, it was suggested that the ideal wheeling rate obtained from marginal costs in (1) be modified using a capital recovery term as follows [10].

$$\begin{aligned}\text{Revenue Reconciled}\\ \text{Wheeling Rate} &= \text{Ideal Wheeling Rate}\\ &\quad + \text{Revenue Reconciliation Adjustment}\end{aligned} \tag{5}$$

However, determining how much of capital should be recovered by the wheeling utility from the wheeling rate, either to recover its network investments or to allocate incentive for future investments, is very difficult. Such revenue reconciliation may also result in large price distortions and gross departures from the basic marginal cost principle.

In this context, a load flow based method in [18] works out the allocation of long run cost of transmission capacity. A distinction is made between incremental and marginal cost based methods in that the former identifies new facilities specifically attributable to specific transmission transactions.

An optimization model is formulated in [21] by taking into account the capacity costs of the transco, as annualized return on invested capital, and variable operating costs. It is seen to recover the capital and operational costs fully for all circuit capacities up to the optimal value.

6 TRANSMISSION OPEN ACCESS AND PRICING MECHANISMS IN VARIOUS COUNTRIES

6.1 United Kingdom

The National Grid Company owns and operates the transmission network and provides the impartial link between competing generators and regional networks- providing the main way to access the market. It is also responsible for the system security and congestion management, which involves procuring and dispatching of ancillary services. Markets for voltage control and reactive power have been created which allows the most cost effective reactive support to be used. Frequency response and black start are handled through bilateral contracts.

With regard to transmission pricing, the tariff is two-part, one fixed and the other variable. The variable component is differentiated on a zone basis, with the basic philosophy being marginal cost-based [22, 23].

The price varies significantly over regions and is based on an *investment cost related policy*, which is similar in concept to a long run marginal cost based price. In effect, the objective is to encourage the generators to invest in those areas where the available capacity is less. This is to avoid congestion in the transmission systems during peak loads.

The south and eastern regions have less generation capacity and thus the generators in these regions pay a negative tariff (*i.e.* receive a payment) for the locational component while in the north since there is more generation but less demand, the customers pay a negligible bill for transmission services.

Therefore, the transmission pricing policy is designed to encourage generation investments in the southern parts of the country, which are predominantly industrial. On the other hand it encourages industrial investments in the northern parts. Those signals are to locate in the system where the existing system can support new generation or be willing to pay a price that reflects the current constraints in the system and the long run costs of reinforcements.

6.2 Chile

The Chilean power system was deregulated in the late seventies and by 1982 introduced a separate transmission pricing policy. Transmission networks have to compulsorily provide an open access to any third party. One of the concerns with the restructuring process and transmission open access has been the trend of gencos and other users of the transmission system users to focus away from transmission planning activities. This has been due to the competitive markets where the gencos primarily seek to increase their profits. To this effect the Chilean system has formulated a tariff structure which recovers costs for investment needs in transmission [1].

The tariff structure is two part- one part is based on marginal cost and the other part, which is fixed, is designed to recover the capital investment costs.

The marginal cost is calculated by dispatching all generation for a single bus system. The marginal cost is then obtained from the dual of the demand balance constraint and other transmission limits. These costs are distributed over the system using loss factors, which thus comprise the first part. Evidently, this component will be very high when transmission constraints are encountered due to the high value of the associated duals of the dispatch model [24, 25, 26].

However, this tariff mechanism, and the revenue surplus from it (net of customer payments and generator costs), does not recover the fixed costs of the line. The second part of the tariff, which is a fixed component, is included to address that aspect.

6.3 Sweden

The Swedish ISO, Svenska Kraftnät introduced the new tariff system for its national transmission network on January 1, '95. The new tariff was formulated to promote competition in the market, by being flexible enough to allow for new types of customers and contracts. Also, it aimed at being simple, open and predictable, so that users could estimate their transmission costs while Svenska Kraftnät could recover its own costs for managing the network, and provide economic and technical signals to the users as regards losses, constraints *etc.* [27].

The tariff scheme is based on a *point-of-connection* principle, which was considered most suitable for open access networks (*Figure 8*).

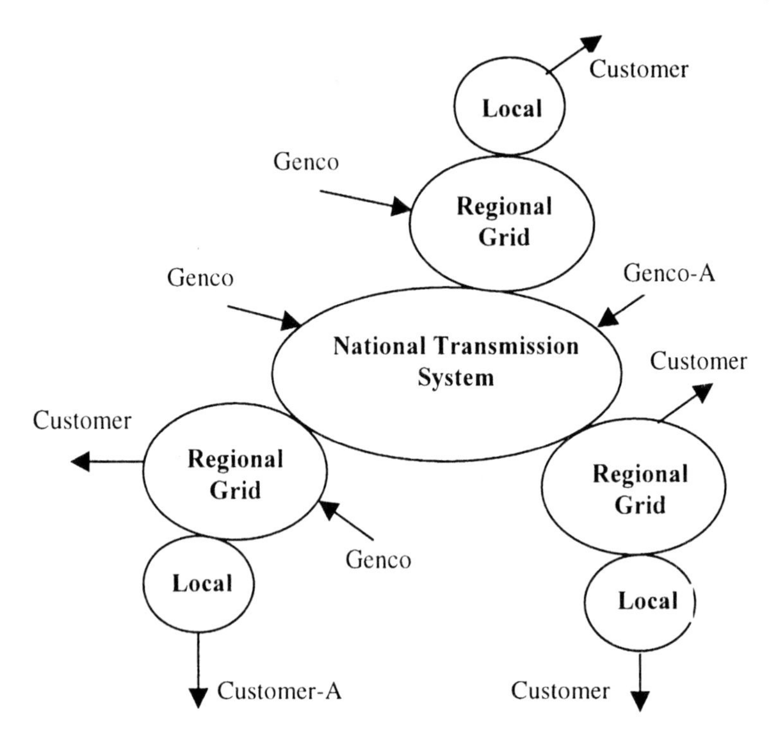

Figure 8. The Swedish network structure: Point of Connection Tariff

The basic principle of this tariff mechanism is that payment made at one point, the point of connection, gives the concerned party (genco, disco, or any customer) an access to the whole network, and thus the whole electricity market. This means that a customer, say customer-A (in *Figure 8*) connected to a local network, pays network fees only to the owner of that network. It can then buy electricity from genco-A, or for that matter, trade electricity with any player within the entire national / regional or local network system.

Similarly, the local-network owner pays network fees to the appropriate regional-network owner to which it is connected. This cost, along with the other costs for running the local network, will be taken into consideration when setting the local-network tariff for customer-A.

Finally, the regional-network owner pays network fees to Svenska Kraftnät for use of the national transmission network. This cost is in turn, taken into account by the regional-network owner in its tariff. Transmission customers are thus largely regional-network owners and generating companies with power stations directly connected to the national network.

This tariff mechanism involving relaying of costs downstream the network is shown in *Figure 9*.

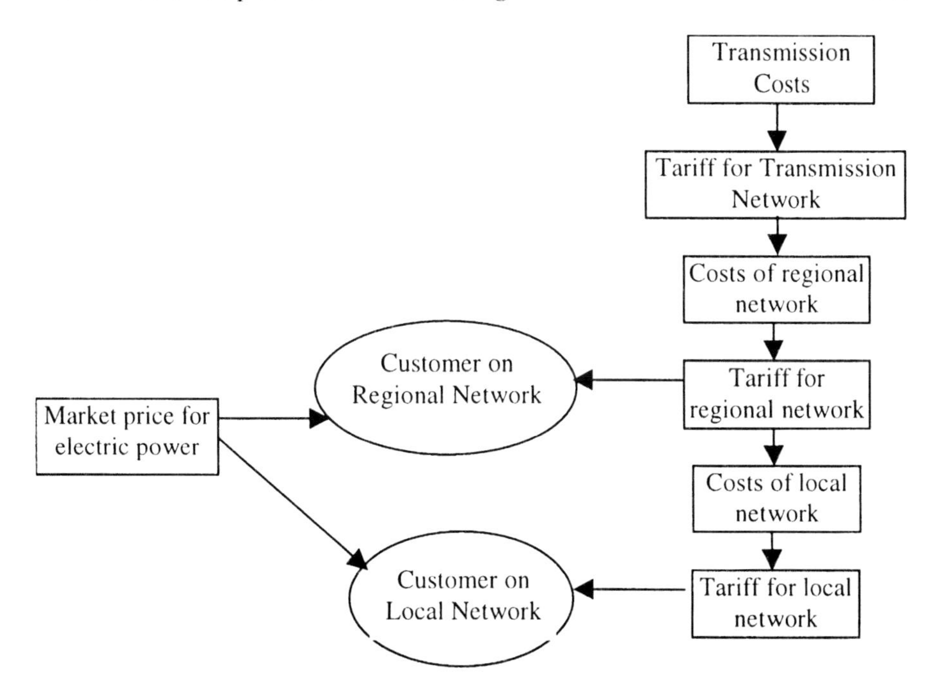

Figure 9. The tariff mechanism in Sweden

The point-of-connection tariff for the national transmission network comprises:

- an annual power fee, intended to provide about half of the revenue
- an energy fee to cover transmission losses, also about half the revenue
- a once-for-all connection fee, which is only applied when connection of a new customer involves considerable costs

The power fee is latitude-dependent (*Figure 10*). This is because of the prevailing power flow in the network from north to south. Generators pay more in the north, where there is a surplus of generation, and less in the south where the load centres and export markets are located. Consumers of power will conversely pay more in the south and less in the north. This is similar to the geographically spaced tariff charged in the UK pool, discussed earlier.

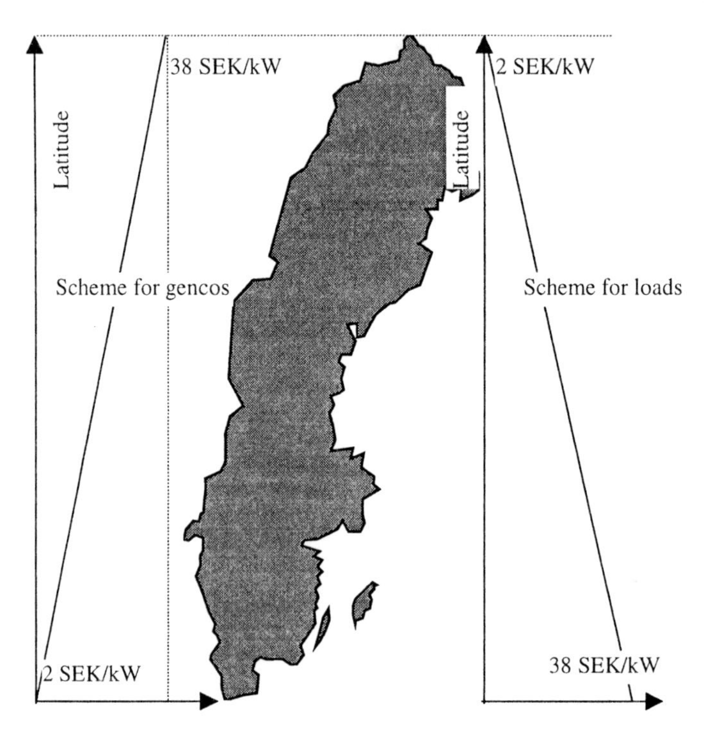

Figure 10. Latitude dependent capacity fee charged in Sweden. There is a similar charge in UK too. This scheme takes into account the long run marginal costs and penalizes the generators in the north and loads in the south. It accounts for an implicit congestion management in the transmission network since bulk power flow in Sweden and UK is from North to South.

The energy fee is calculated as the product of the appropriate marginal-loss coefficient for the connection point, the energy input or output at this point, and the current power price. Marginal-loss coefficients for high and low load, weekday and other times, are calculated for each connection point in the network and can be positive or negative.

7 DEVELOPMENTS IN INTERNATIONAL TRANSMISSION PRICING IN EUROPE

The Transmission System Operators (TSO) in Europe are expected to adopt a new pricing scheme for their international transactions. For example, the transaction between an independent genco located in one country selling power to a customer located in a different country. In such a case, all TSO involved in the transactions *i.e.* the TSO of the exporting country, the TSO of the importing country, and all intermediate TSO over whose network power flows are expected to take place, will receive payments from the transacting parties. However, the payment received by the TSO will depend upon the role performed by it in the transaction and accordingly, they would have to notify their tariff as per their costs involved in each activity of transaction. The three components that have been identified for payment, and will be applicable as per the role of the TSO in a transaction, are as follows [28]:

Component-G: This is payable to the TSO in which the genco is located for providing the access to the network (either directly or through downstream networks) for the export. This component comprises a charge on a per-unit basis (per kW or per kWh) for the cost incurred in allowing it to reliably access the TSO grid.

Component-L: This is payable to the TSO in whose area the customer or the load is located and comprises a charge on per unit basis representing costs incurred in allowing the customer to reliably access the grid.

Component-T: This is payable too all those TSO involved in the transaction by way of providing for the power flow between the genco and the customer. This also includes those TSO over whose systems any parallel flows take place due to the transactions. The charge comprises a per unit charge representing the cost of transmission on a per unit basis (per kW or per kWh, and can be distance dependent) incurred by the intermediate TSO.

Each individual TSO is required to disaggregate its total transmission costs into the above three components for international transactions.

7.1 Example

Consider that a genco in country 'A' contracts for power delivery of 100 MW to a customer in country 'D' as shown in *Figure 11.* The interconnections between the countries are as shown, and it is found from load flow studies that the power flow (without considering losses) will actually take place over the interconnections as denoted by the arrows.

The draft rules in [28] specify that the transaction charges should be formulated based on physical power flows obtained from realistic power flow models or on the basis of actual power flow measured on the tie-lines.

Let us assume that the four TSO involved in the above transaction have pre-decided tariff components for G, T and L (arbitrarily chosen) as follows:

Table 6. Arbitrarily chosen tariff components for the example case

TSO	G, in $/MW	L, in $/MW	T, in $/MW
A	4	2	5
B	5	6	2
C	9	1	7
D	8	5	3

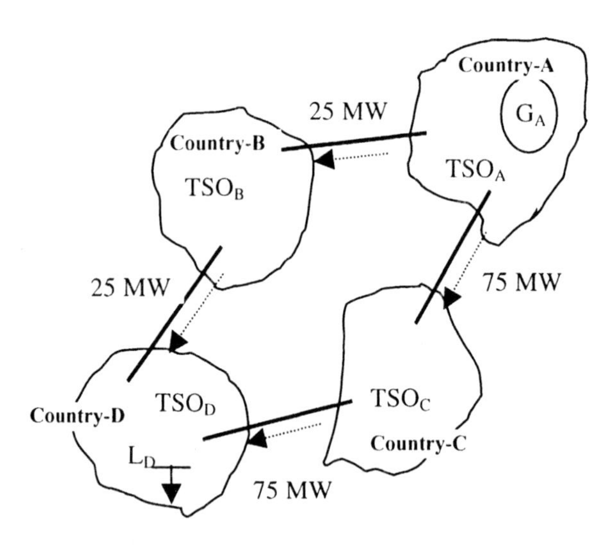

Figure 11. The proposed transmission pricing mechanism to be adopted in Europe for inter-country power trades.

For the power flows taking place as shown in *Figure 11*, and the tariff components G, L and T as in *Table 6*, the four TSO involved in the 100MW transaction will receive payments as follows:

Table 7. Payment received by a TSO as per the different components of tariff

TSO	G, in $	L, in $	T, in $	Total receipts, in $
A	4*100 = 400	-	5*100 = 500	900
B	-	-	2*25 = 50	50
C	-	-	7*75 = 525	525
D	-	5*100 = 500	3*100 = 300	800

The total transmission payment involved in the transaction of 100MW of power from country 'A' to country 'D' is thus, $2275 which is to be paid by the genco or the customer in country 'D' as per their mutual agreement.

8 SECURITY MANAGEMENT IN DEREGULATED ENVIRONMENT

The ISO is faced with the difficult task of providing security constrained transmission services in an open-market environment that is fair and equitable. In this environment of free markets, the transmission network is liable to be heavily stressed due to the various trades and transactions. It is also to be noted that these networks were originally designed to accommodate transactions following only certain load / generation patterns at best. In deregulation the generation pattern resulting from the market activities can be quite different from the traditional one, possibly worsening the security margins.

Furthermore, since any generator can sell all or part of its output to multiple buyers located anywhere within the network, and there could be a large number of simultaneous transactions taking place, accountability and traceability based on transactions become crucial issues. In a similar context, planners are confronted with important issues such as how to identify potential bottlenecks in the network and the most suitable and profitable expansion plan while taking into account the reliability of the system and all existing and future market activities.

The basic problem, therefore, that has emerged from deregulation is that-to maintain system security at a desired level, how can the ISO evaluate the transactions that don't violate system security. The biggest challenge is that, to some extent, every transaction has an effect on all other transactions. For example, by changing one particular transaction, some other transactions may also have to be modified in order to meet the power balance, generation limits and security requirements.

In pool type of markets security constraints are often incorporated within the market settlement process. For example, in the New York (NY) and PJM (Pennsylvania, Jersey, Maryland) power pools, the ISO carries out a security constrained economic dispatch simulation based on energy price offers, start-up cost offers, and minimum and maximum generation levels [29]. In the case of PJM, the security constraints are subsequently input to a unit commitment model by the ISO to obtain generation and commitment schedules which abide by security constraints. In the New Zealand Electricity Market (NZEM) various security constraints are simultaneously incorporated within the market clearing mechanism [30]. The scheduling, pricing and dispatch (SPD) model takes into account branch flow limits, generator ramp rates, fast acting reserves (6 seconds), sustained reserve, (acting within 60 seconds), and various restrictions on clearance of spinning reserves and reserve capacity vis-à-vis the amount of power generation bids cleared. The reserve scheduling also includes scheduling of interruptible loads. *Table 8* provides a summary of how security constraints are handled in some of the deregulated electricity markets.

Table 8. Security Management Practices in Some Electricity Markets

Market	How are transmission / security constraints handled
UK	Transmission limits are modeled as import / export constraints in the centralized market settlement model
NY	Security constrained economic dispatch considering line limits on specific lines using a dc power flow simulation
PJM	Security constrained economic dispatch followed by unit commitment that receives security constraints as inputs for the unit commitment model
NZEM	The market clearing model incorporates linear power flow representation with line flow limits, constraints on branch group are modeled as linear functions of power flow
Nordic Countries	Transmission limits are enforced using the price area segregation method in Norway and counter-trading mechanisms in Sweden.

All the above electricity markets, except for the Nordic countries, operate as a centralized power pool and hence the ISO in all cases incorporates security constraints within the market settlement process, as we have seen in *Table 8.*

In the case of Nordic countries however, the ISO has neither a role in the market settlement or in the bilateral contracts settlement. Therefore, system security is handled by the ISO outside the market settlement process, through the balance market and counter-trades. It is also possible to represent security constraints within an ISO optimization problem for bilateral contract markets, as shown in [31, 32 and 33]. The ISO's objective can be represented as the minimization of a penalty function of the type,

$$J = \sum_{i}\sum_{j}\left(T_{i,j} - T_{i,j}^{Allowable}\right)^2 \tag{6}$$

In (6) T is the bilaterally contracted transaction matrix between a supplier and a customer and is known *a priori*, while $T^{Allowable}$ is the allowable transaction decision variable, determined by the ISO considering system security aspects, and is subjected to constraints on upper limit on allowable transactions, T^{Max}.

$$0 \le T_{i,j}^{Allowable} \le T_{i,j}^{Max} \tag{7}$$

The power flow equations on lines can be modeled using a dc power flow representation or a full ac OPF model.

In [32] a set of probabilistic indices have been defined to assess the capability of the transmission system, with multiple transactions in random combinations and magnitudes taking place in the system. These indices help the operator and planner to determine the state of system security vis-à-vis the bilateral transactions, and also the needed reinforcements in the network.

8.1 Scheduling of Spinning Reserves

The scheduling of spinning reserves is very important in the daily operation of power systems. In the classical vertically integrated environment, the most common way to represent this is by including a reserve margin equal to the size of the largest generating unit or some percentage of the peak load in the unit commitment program. Such a method is still possible in a deregulated environment, where the pool is the central dispatcher and is responsible for unit commitment decisions. The only difference being that the pool ISO takes into account the price bids for reserves instead of generating cost.

Mathematically, the optimization problem can be described as the minimization of,

$$J = \sum_{k=1}^{T} \sum_{i=1}^{N} \left[C_i(P_{i,k}) + UST_{i,k} \cdot ST_{i,k} \right] \tag{8}$$

In the above equation, P is the generation from unit *i* at time *k*, UST is the unit start-up decision variable (1 = ON, 0 = OFF), C is the genco's price bid and ST is the genco's start-up price bid. As discussed in Chapter-3, the market will be settled at that point, where the total scheduled energy meets the demand forecast while taking into account the spinning reserve constraint. Due to the introduction of unit ramp up limits, the spinning reserve amount contributed by each unit must be modified as follows [34, 35]:

$$r_{i,k} = \min\left\{ P_i^{Max} - P_{i,k} ; \tau \cdot W_{i,k} \right\} \tag{9}$$

$$\sum_{i=1}^{N} r_{i,k} W_{i,k} \geq R_k ; \qquad k = 1, \ldots\ldots, T \tag{10}$$

Where *r* is the unit reserve constraint, R is the spinning reserve requirement and τ is the reserve response time (about 10 minutes).

8.2 Interruptible Load Options for Security Management

Interruptible load schemes offer the customers a range of rate-reliability choice. In this option, the customer signs an interruptible load contract with the utility to reduce its demand as and when requested. The utility benefits from this by way of a reduction in its peak load, which helps it to restore its quality of service (voltage profile, etc.) and ensure reliability. The customer entering such a contract, benefits from reduction in its energy costs and from other incentives provided by the utility.

Customers, who choose to buy interruptible energy, can do so by communicating the secure energy level they want, so that all usage above that level is at the interruptible price. If the utility experiences short-term emergency operating conditions, the reduction required from each customer is computed and communicated to them. Thus, interruptible loads can aid the utility in maintaining an on-line reserve (which could be activated fast enough, possibly within one hour). It could also assist in maintaining voltage profile, providing real and reactive power relief, etc.

In the context of deregulation, a market for interruptible load (*callable forward*) contract, which is continuously tradable until the time of use, is a feasible option [36]. Since the contract specifies a fixed payment for interruption, the utility has no obligation to provide the customer a certain availability of service, frequency and duration of interruption, etc. Another form of interruptible load used in UK involves large industrial customers bidding into the pool directly on their ability to reduce load, referred to as demand-side bidding (DSB) [37, 38].

8.2.1 International Practices in Interruptible Load Management

8.2.1.1 Demand-side Bidding Mechanism in UK

Within the trading arrangements of the UK power pool, demand-side bidding was introduced in December 1993 and has since then operated as a demand reduction scheme. In this way, the demand-side bidders are deemed to have a more beneficial effect by reducing demand by a pre-defined amount rather than by an unknown amount. The participants must offer at least 10 MW of their load for curtailment and have a potential for 50 GWh demand reduction over a year[2]. Demand-side bidders are expected to abide by the demand reduction schedules, or if no schedule is received, when the system marginal price is equal or higher than the bid price of the relevant reducible demand.

[2] This implies that the demand-side bidders should be prepared for about 1000 hours of curtailment per year

Following the payment structure discussed in Section-2.1.1 of Chapter-3, the payment structure for demand-side bidding is structured as follows:

- Demand-side bidders pay at *Pool Selling Price* for all demand actually taken, independent of whether it was offered as being reducible.
- Demand-side bidders are paid an *Availability Payment*, when there is a value, for all demand offered available for reduction, that is not scheduled in the unconstrained schedule.

The intent behind the scheme is to schedule, when cost-efficient, any demand reduction submitted by participants as available for reduction in a similar manner as generating units. The scheme is implemented as follows:

The demand-side bidders bid in their fully expected demand for each half-hour of the next day, offered reducible availability, which is the demand available for reduction and the market price above which, the demand will be reduced. The pool operator resolves the market incorporating the demand bids that are scheduled in the unconstrained market settlement in the same manner as scheduling a generating unit. Within one hour of the publication of the system marginal price, the demand-side bidder will receive notification of demand reduction scheduled for the next day. In the event that there is no demand scheduled for reduction, then whenever the value of the marginal price equals or exceeds the bid price, participants are required to reduce demand [37, 38].

8.2.1.2 New York ISO (NYISO)

There is a provision for customers to offer interruptible load service to a Load Serving Entity (LSE) within NYISO and thereby provide additional operating reserve to the later. They may enter into contracts with LSE for compensation. But in order to bid in the day-ahead or operating reserve market, customers must contract their interruptible load with NYISO directly thereby allowing direct control, monitoring and billing by the latter. The bids must be larger than 1MW, the response time 10 minutes and the duration can be up to 1 hour. Interruptible loads are classified into several types:

- *With non-price capped fixed energy:* Load that schedules non-price sensitive energy (*i.e.* a fixed MWh level with no price cap), and then offers to interrupt that load to reduce the demand
- *With price-capped energy:* Load that schedules day-ahead price-sensitive energy, and then offers to interrupt that load to reduce the demand.

There is a provision for 10-minute and 30-minute spinning reserve market in NYISO wherein interruptible and / or dispatchable load resources located within the NYISO and synchronized to the system can offer to participate. In such cases, they would need to respond to the ISO instructions for load curtailment within the 10-minute or 30-minute time frame, as applicable. The bids for these markets can be for 2 MW or 1 MW of synchronized load at each hour and the NYISO schedules for both 2 MW load and 1 MW 10-minute spinning reserve for each hour. The 2 MW loads are paid for each hour at day-ahead energy price while the 1 MW loads are paid the 10-minute spinning reserve market-clearing price for each hour [39, 40].

8.2.1.3 Demand Relief Program of California ISO (Cal-ISO)

Cal-ISO has initiated a demand relief program in which the customer signs a contract with the ISO for its demand reduction. The ISO implements the program as a means of providing incentives to induce customers to reduce their demand during times of resource shortage. The payments are based on monthly capacity reservation, which is preset by the ISO, and then there are payments for the energy actually delivered. The customer must be able to reduce at least 1 MW of load demand. The ISO and contracted load enters into the contract, where loads without back-up generation are called first, for interruption when emergency reserves fall below 5%. These loads can be curtailed up to four hours, in multiple blocks, and need to respond to curtailment signal within thirty minutes. Loads having a back-up generation source are called second for interruption, when reserves fall below 2%. These loads need to respond within fifteen minutes [41].

8.2.1.4 Alberta Power Pool

The Alberta power pool in Canada has in place a curtailable-load program so as to enhance the system security of the Alberta Interconnected Electric System operations. Customers willing to participate in the program need to offer at least 1MW of their load as curtailable load. There is also a requirement for time-of-use metering equipment and the customer's ability to receive dispatch instructions from the pool when required.

Customers need to submit their bids in terms of the curtailable MW on offer and the associated price. Based on the bids received by the pool, the pool operator determines the contracted curtailable customers, by classifying them either as Type-1 or Type-2. The Type-1 contracts are for a one-month period and require the customer to reduce its load for up to a four-hour duration, with the response being within one hour, and with a maximum number of five interruption requests per month. The financial compensation is based on a fixed price per MW per month, independent of the number of interruptions requested. The Type-2 contracts are somewhat short-term and are for a one-week period, requiring the customer to reduce its load within one hour up to a four-hour duration. Financial compensation is based on a price per MWh basis and is paid only when curtailment is activated [42].

Bids are ranked in order of the prices, from highest to lowest. The pool operator selects the bids starting from the lowest bid price till it meets its curtailment quantity requirements.

8.2.1.5 NEMMCO Australia

The National Electricity Market Management Company (NEMMCO), which is the ISO in Australia, allows for demand side bidding in the market. However the rules and codes are not very attractive to the customers. So far, only a few large customers and pump-storage hydro power plants are participating. Customers can register as scheduled loads and can submit their dispatch bids to the NEMMCO. Both generators and customers are centrally dispatched.

The dispatch bid can be specified so as to increase or decrease the load if the price is below or above the pre-specified level. The Australian National Electricity Code Administration is taking initiatives to change the rules to introduce more attractive arrangements for demand side bidding [43].

8.2.2 Optimal Incentives for Interruptible Load Contracts

The success of an interruptible load program often depends on the incentive rate structure being offered by the utility. The utility offering the interruptible load program therefore needs devise appropriate incentive rates that would induce the best customer response. These should also be sustainable to the utility considering its profit objectives. The incentives may change from hour to hour depending on the utility's reserve availability and load serving capability. Thus if the system is operating near its capacity margin the incentives could be high, while with adequate operating reserve, they would be low. The incentives may also vary with customer groups, because of their difference in quality of load. For example, interruption of low power factor agricultural load will relieve the reactive power demand in the system more and thus aid in voltage profile improvement. Further, the incentive could also vary with location of the customer. So, for an interruptible load scheme to succeed, the utility needs some idea on the responsiveness of a customer group to the scheme being offered.

8.2.2.1 Working of a Real Time Interruptible Load Scheme

Although most interruptible load contracts are specified well in advance, present-day energy transactions allow for much more frequent updates such as one-hour spot-price, calculated based on system operating conditions and forecasts of how much interruptible energy will be purchased by customers. A real time interruptible load scheme is discussed below based on an hour-to-hour determination of incentive rates through the use of an OPF [44].

In this scheme the utility should announce on an hour-to-hour basis, the previous hour's spot-price difference (SPD) between contracted interruptible load (CIL) and rest of the loads. The customer choosing to buy interruptible load can communicate the secure energy level it wants, so that all usage above that level is at the rate of interruptible tariff. If the utility experiences short-term emergency operating conditions or even otherwise, the reduction required from each customer is computed and communicated to them (*Figure 12*).

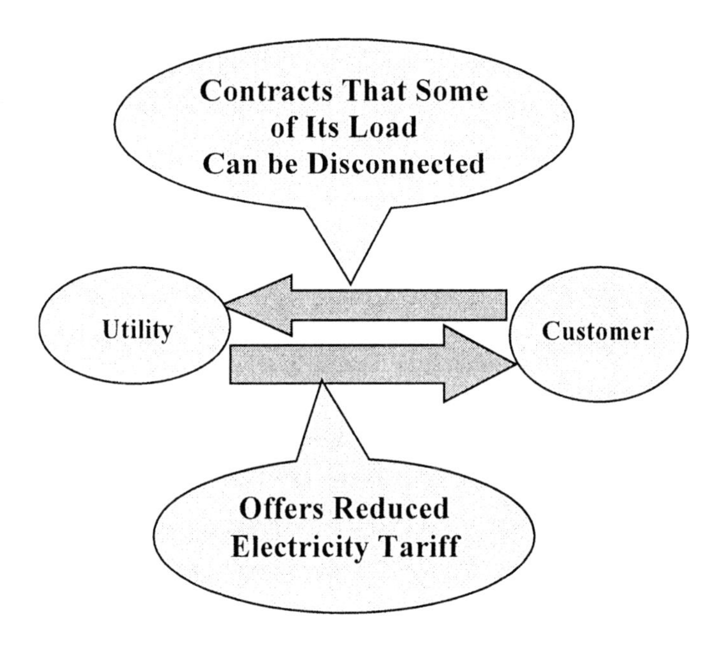

Figure 12. Working Scheme of the Interruptible Load Program

If the SPD was high during the past hour, more customers are likely to opt for ITM during the next hour, *i.e. CIL is sensitive to SPD*. On the other hand, higher the incentive given by the utility in an hour, lesser will be its willingness to interrupt, *i.e.*, utility affected *load interruption is incentive sensitive*. The optimal incentive can be determined from the equilibrium of the two sensitivities *i.e.*, when the utility sensitivity $\eta^{Utility}$ (11) is equal to the customer's sensitivity $\eta^{Customer}$ (12) to interruptible tariff.

$$\eta_{i,Type}^{Utility,k} = \frac{\Delta PD_{i,Type}^{k}}{\Delta \rho_{i}^{k}} \tag{11}$$

$$\eta_{i,Type}^{Customer,k} = \frac{CIL_{i,Type}^{k}}{SPD_{i}^{k-1}} \tag{12}$$

$$\text{Where, } \mathrm{SPD}_i^{k-1} = \rho_{\mathrm{BASE},i}^{k-1} - \rho_{\mathrm{ITM},i}^{k-1} \tag{13}$$

In the above equations, *k* is the index for hour, *i* for bus and *Type*, for type of customer (industrial, commercial, domestic, *etc.*). ΔPD denotes the amount of interruption and $\Delta\rho$ is the incentive rate to customers opting for ITM. SPD is the spot price difference between the normal price (ρ_{BASE}) applicable to general customers and the reduced price (ρ_{ITM}) applicable to ITM customers, and this is determined for the previous hour.

8.2.2.2 Determining the Optimal Interruptible Tariff

One factor that will determine the success of the real-time interruptible load program, is the customer response to the hourly SPD prevailing. However, estimating the customer response is very subjective and can vary with the type of customer, its location, *etc.* and a survey would be an effective method to determine the expected customer response to the incentive rate being designed. In the absence of a survey, certain mathematical formulations for customer response could aid in understanding the customer behavior to the interruptible load program.

It can be assumed that in real-time operation of the interruptible load program, the customer response at the kth hour is influenced by the SPD prevailing at the (k-1)th hour. The degree of influence is governed by a *sensitivity index* (SI), specific to customer type. SI is the customer's perception of ITM and indicates, how much of the load it is likely to contract as interruptible, per $ of its bill savings. That can explain why an industrial and an agricultural customer would respond differently to the same SPD. A typical model for customer response function is given in (14) to represent the customer response to interruptible tariffs.

$$\mathrm{CR}_{i,\mathrm{Type}}^{k} = 1 - e^{-\mathrm{SPD}_i^{k-1} \cdot \mathrm{SI}_{\mathrm{Type}}} \tag{14}$$

CR is the customer response to the information made available every hour, on spot price difference of the previous hour. In (14) it is assumed that CR varies exponentially with SPD and theoretically would tend to one (*i.e.* 100% of the load is contracted for interruption), as SPD tends to infinity. However, in realistic utility-customer transactions, it is not expected that more than a certain percentage of the total load in a particular category will opt for interruption. Thus, an upper limit on the customer response can safely be assumed. The CIL can then be determined as follows,

$$\mathrm{CIL}_{i,\mathrm{Type}}^{k} = \mathrm{PDem}_{i,\mathrm{Type}}^{k} \cdot \mathrm{CR}_{i,\mathrm{Type}}^{k} \cdot \mathrm{CF}^{k} \tag{15}$$

In (15) CF is an adaptive parameter that needs to be updated every hour based on the past hour's ratio of actual interruption and CIL. *PDem* is the demand PD, disaggregated at a bus over different *Type* of customers. The above constraint restricts disproportionate interruptible load contracts. For example, if CIL is large compared to the actual interruption during an hour, CF will provide a proportional reduction in CIL during the current hour. The working scheme of the ITM can be outlined by the following steps.

- *Get SPD for $(k\text{-}1)^{th}$ hour from the base-case and modified OPF models*
- *Determine CR for the kth hour*
- *Determine CIL for k^{th} hour and hence $\eta^{Customer}$*
- *Solve modified OPF that includes the equilibrium condition, $\eta^{Customer} = \eta^{Utility}$, to determine $\Delta\rho$ for k^{th} hour*
- *Evaluate interruptible tariff (TITM) for k^{th} hour as follows:*

$$\mathrm{TITM}_i^k = \rho_{\mathrm{BASE,i}}^k - \Delta\rho_i^k \tag{16}$$

$\Delta\rho$ is the optimal tariff incentive which is determined from the model, while TITM is the final tariff charged to the customers. TITM thus evolved is available as spatially differentiated spot price. $\rho^k_{\mathrm{BASE,i}}$ is the utility's spot-price obtained from a base-OPF without ITM.

8.2.2.3 OPF Incorporating Utility-Customer Interactions

The modified OPF model incorporates the utility-customer interactions in an ITM and inter-temporal ramp-rate checks that constrain large variations in generation, following the hour to hour variations in load. The objective function is augmented to include a term representing the utility's financial returns.

Utility Financial Returns:

If the utility charges its customers at marginal cost prevailing at a load center, $\rho^k_{\mathrm{BASE,i}}$, its hourly profit, PR, from sale of energy can be given by,

$$\mathrm{PR}_i^k = \rho_{\mathrm{BASE,i}}^k \cdot \mathrm{PD}_i^k - \mathrm{C}_i(\mathrm{P}_i^k) \tag{17}$$

In (17) C(P) is the generation cost incurred by the utility to meet the demand. With ITM, the utility charges the CIL portion at TITM rate, and rest of the load, at $\rho^k_{\mathrm{BASE,i}}$. The revised profit calculations can thus be given as,

$$PR_i^k = \rho_{BASE,i}^k \cdot \sum_{Type} \left(PDem_{i,Type}^k - CIL_{i,Type}^k \right) + TITM_{i,Type}^k \cdot CIL_{i,Type}^k - C_i(P_i^k) \tag{18}$$

Algebraic manipulation of (18) shows that the utility's reduction in revenue earnings in the ITM case, results from an additional term, $\Delta\rho^k_{i,Type}.CIL^k_{i,Type}$. The utility's objective will be to keep this as low as possible. Note that there will be no reduction in profit, though.

The Model:

Objective Function: In the short-term, the utility's objective is to minimize its hourly generation cost, operational costs incurred in providing reactive power support at load buses, and the revenue lost in providing incentives to customers participating in the ITM. Therefore, the objective function for the utility can be expressed as an hourly cost,

$$J^k = \sum_{i} C_i(PG_i) + \sum_{i} QC_i^k CQC_i + \sum_{i} \sum_{Type} \Delta\rho_i^k CIL_{i,Type}^k \tag{19}$$

QC is the reactive support at a bus and CQC is the associated short-term marginal cost of that reactive power support. The first term in (19) denotes the total cost of generation, the second term denotes the total cost of providing reactive power support and the third term denotes the revenue lost in providing for incentives to interruptible load customers.

Load Flow Equations: These are modified to include the interruptible load, ΔPD, requested by the utility from those customers participating in the ITM.

$$P_i^k - PD_i^k + \sum_{Type} \Delta PD_{i,Type}^k = \sum_j V_i^k \cdot V_j^k \cdot Y_{ij} \cdot \cos(\theta_{ij} + \delta_j^k - \delta_i^k) \tag{20}$$

$$Q_i^k - QD_i^k + QC_i^k + \sum_{Type} \Delta QD_{i,Type}^k = -\sum_j V_i^k \cdot V_j^k \cdot Y_{ij} \cdot \sin(\theta_{ij} + \delta_j^k - \delta_i^k) \tag{21}$$

$$\sum_{Type} \Delta QD_{i,Type}^k = \tan\left\{\cos^{-1}\left(PF_{Type}\right)\right\} \cdot \Delta PD_{i,Type}^k \tag{22}$$

Q is the reactive power scheduled by the utility from generating units and ΔQD is the relief in reactive power associated with the real power interruption ΔPD. The relief received by the utility depends on the type of customer being interrupted, because of the differences in power factors (PF) of various types of customer loads. V, δ, Y and θ are the bus voltages, voltage angles, line admittance and the angle associated with the line admittance respectively.

Utility-Customer Interactions:

- Interruption affected by the utility is governed by the equilibrium condition, $\eta^{Customer} = \eta^{Utility}$ and (11).

$$\Delta PD_{i,Type}^{k} = \eta_{i,Type}^{Customer,k} \cdot \Delta\rho^{k} \tag{23}$$

- Upper limit on load interruption will be limited by utility-customer contract, the CIL.

$$\Delta PD_{i,Type}^{k} \leq CIL_{i,Type}^{k} \tag{24}$$

- Upper limit on CR, beyond which, whatever the SPD, the customer response will not increase.

$$CR_{i,Type}^{k} \leq CR_{i,Type}^{Max} \tag{25}$$

- Upper limit on incentive that the utility can sustain, without compromising on its financial returns. It can be assumed limited by a fraction a_o of the base spot price prevailing at an hour.

$$\Delta\rho_{i}^{k} \leq a_{o} \cdot \rho_{BASE,i}^{k} \tag{26}$$

- The utility could tend to contract more interruptible load than the actual interruption it asks for. That would lead to unwarranted payment of incentives to the customers. The adaptive contract factor CF revises the CIL based on the amount by which, during the past hour, the CIL exceeded the actual interruption.

$$CF^{k} = \frac{\sum_{i}\sum_{Type} \Delta PD_{i,Type}^{k-1}}{\sum_{i}\sum_{Type} CIL_{i,Type}^{k-1}} \tag{27}$$

- Other constraints in utility-customer interactions need to be imposed.

$$\Delta\rho_i^k = 0, \quad \forall\ CR_{i,Type}^k = 0, \ \text{or}\ \eta_{i,Type}^{Customer,k} = 0 \tag{28}$$

Other Constraints

- Upper and Lower Limits on Generation
- Upper and Lower Limits on Bus Voltages
- Upper and Lower Limits on Reactive Support at Load Buses
- Hourly ramp-rate check

  The dispatch must abide by the unit's ramp-up (or backing down) limits. These would depend on the previous hour's dispatch status and the unit's ramp-rate characteristic. If the generating limits are exceeded, the generation level is set at the limits.

In [45] a comprehensive pricing formulation for interruptible loads and assignment of power pool reserves have been worked out. It is shown that optimal pricing mechanisms do exist, and these invoke customer participation in a socially optimum manner to aid in system operation and provide for system security. Consequently, it is shown in [46] that system security can be maintained in an operating environment where all participants (including those on the supply-side and those on the demand-side) seek to optimize their own benefits through pricing mechanisms. A generalized model for inclusion of security constraints in competitive markets is developed in [47] where the prices are determined considering customer demand-price elasticity when.

9 CONGESTION MANAGEMENT IN DEREGULATION

In the vertically integrated utility structure, all entities such as generation, transmission and distribution are within the domain of a central energy management system. Generation is dispatched in order to achieve the system least cost operation. In such systems, congestion management is usually taken care of by determining the optimal dispatch solution using a model similar to the optimal power flow or the security constrained economic dispatch problem.

This effectively means that a generation pattern is determined such that the power flow limits on the transmission lines are not exceeded. The presence of transmission line capacity constraints in these scheduling programs leads to higher marginal cost and reduced revenue for the utility, that in turn act as a signal to the utility. A persistent congestion problem is an indication to install new generation capacity or to build additional transmission facilities [48].

Apart from alleviating transmission congestion through the dispatch and scheduling process, other methods such as operation of transformer taps, outage of the congested line and operation of FACTS devices, *etc*. are also available with the ISO. In these discussions however, discuss congestion management methods through the application of economic instruments, market based methods and other pricing tools.

9.1 Economic Instruments for Handling Congestion

Lot of work has been reported in the area of congestion management in deregulated power systems using mathematical models that incorporate pricing tools. In [49] a detailed analysis has been carried out using pricing as an instrument for congestion management within an optimal power flow framework. It is concluded that the ISO need not know the marginal cost of the suppliers in order to manage congestion. They can infer the required parameters from an observation of the market.

In [50, 51], a market structure where both pool and bilateral contracts co-exist has been considered, and a congestion management scheme developed by prioritization of the transactions and curtailment of certain transactions. A parameter termed *willingness-to-pay-to-avoid-curtailment* has been discussed which can be an effecting instrument in determining the transaction curtailment strategies. In addition, there can be a prioritization of transactions, for example, pool trades being accorded a higher priority over bilateral trades; or the ISO seeking to protect the bilateral contracted transactions, as far as possible, while curtailing pool based transactions.

In [52] a general modeling framework has been developed that helps in analyzing power transactions in deregulated markets subject to security and congestion constraints. It is suggested that economic values associated with transactions need to be disregarded during times of critical conditions in the system and transaction rescheduling undertaken, so as to arrive at a point within the security region that is closest to the scheduled transactions.

In the following section we discuss the method of market splitting to alleviate transmission congestion. The basic principle of this method lies in sending price signals to generator that are higher or lower than the marginal cost (or system price, as the case may be) that would result in a change in their generation pattern. The same principle can be applied to the customer-side, and customer demand elasticity to price signals can help in congestion management, as demonstrated in [47].

9.1.1 Market Splitting[3]

In this method, the market-clearing price (ρ_M) is first calculated without considering any transmission constraints. This unconstrained price is applicable to all participants, both supply-side and demand-side, and irrespective of their location, if and only if no transmission congestion arises from the market settlement.

However, if the market operator expects that due to the market settlement, the resultant power flow may introduce transmission constraints in the system, the system price no longer remains the same everywhere. In such a situation, the market is split into different bid areas and the area-prices (ρ_{AREA}) are calculated for each bid-area using a *capacity fee*.

The market price is reduced in the generation-surplus area and increased in the generation-deficit area until the power transfer need between the two areas is brought within the transmission capacity limit. Such a differential pricing instrument provides a signal to purchase more and sell less in the low price area, and sell more and purchase less in the high price area.

To compute the area price in the low-price area (surplus area), the seller is debited and the buyer is credited the capacity fee, while in the high-price area (deficit area) the seller is credited and the buyer is debited the capacity fee. In effect, this provides an incentive to those who aid in relieving the system overload and penalizes those who worsen it.

Thus, the area price in the low price area (generation surplus area) is given by:

$$\begin{aligned} \rho_{AREA}^{Sell} &= \rho_M - \rho_{CAP} \\ \rho_{AREA}^{Buy} &= \rho_M - \rho_{CAP} \end{aligned} \qquad (29)$$

The above price structure penalizes the seller in the surplus area with a price reduction by 'ρ_{CAP}', and allowing a lesser price than the market price, while providing an incentive to the buyer, who pays a reduced price.

On the other hand, the area price in the high price area (generation deficit area) is given by:

$$\begin{aligned} \rho_{AREA}^{Sell} &= \rho_M + \rho_{CAP} \\ \rho_{AREA}^{Buy} &= \rho_M + \rho_{CAP} \end{aligned} \qquad (30)$$

[3] This way of splitting the market into different price areas in order to remove congestion is practiced in Norway. The Swedish and Finnish markets always operate as one price areas while Denmark has two price areas.

This pricing structure penalizes the buyer in the high price area by increasing the price above market price with the capacity fee of 'ρ_{CAP}' while it provides an incentive to the seller, who now sells at a price above the market price.

Consider a double auction power pool wherein both gencos and customers bid for power sell and purchase respectively. As we mentioned earlier in this section, the market is first settled assuming that there are no transmission bottlenecks. With this assumption, this is similar to the double auction pool formulation discussed in Chapter-3, Section-3.1.2.1. We briefly discuss this unconstrained model again, for the sake of convenience and to correlate it with the congestion management model to be discussed next.

Let us assume a power pool comprises participants located at different geographical areas, and the areas have transmission interconnections between them. The pool receives supply-bids from gencos for sell and demand-bids from customers for purchase, from each respective area *i*. The generation bids are assumed to be of the form,

$$PG_i(\rho) = b_{o_i} \cdot \rho_M + c_{o_i} \tag{31}$$

The demand bids are of the form,

$$PD_i(\rho) = b_{1_i} \cdot \rho_M + c_{1_i} \tag{32}$$

In the above, ρ_M is the market price (in \$/MWh), which is to be determined. Coefficients b_o and b_1 are price-dependent quantity bids (in MW per \$/MWh), while c_o and c_1 (in MW) are quantity bids, independent of the market price. In the unconstrained case, the market is settled where the aggregated generation bid curve intersects the aggregated demand-bid curve, *i.e.*, simply by equating the total generation with the total demand, *i.e.*,

$$\sum_{i=1}^{N} PG_i(\rho_M) - \sum_{i=1}^{N} PD_i(\rho_M) = 0 \tag{33}$$

The solution of the above will yield the unconstrained market price, the amount of generation and demand cleared in the market and the power transactions that will take place between the different areas. Net transfer from an area can be determined after the market is settled using the basic demand balance equivalence for each area separately.

If these net transfers over the transmission links are higher than the capacity earmarked by the pool operator, the whole settlement is re-worked by incorporating the transmission constraints in the market settlement, which is now carried out individually for each area, as follows:

$$PG_i(\rho_M) - PD_i(\rho_M) + \sum_{j}^{N} T_{i,j} = 0 \qquad \text{for all } i \tag{34}$$

Subject to the constraint on transmission of power,

$$T_{i,j} = T_{i,j}^{Max} \tag{35}$$

Note that in the above, the transmission constraints have been formulated as linear flow constraints. More complex representations than the linear flow, such as the dc load flow or ac load flow can also be used. The above discussion is now explained with the help of an example, which is an extended version of the example described in [53 and 54].

9.1.1.1 Example

Consider a four-area power pool where the market operator receives a sell-bid and a buy-bid from each area, as given in *Table 9.* The market operator first settles the market assuming no transmission congestion.

Table 9. Information received by the market operator on bids from gencos and customers

Area	Sell Bid		Buy Bid	
	b_0, MW/($/MWh)	c_0, MW	b_1, MW/($/MWh)	c_1, MW
A	10.2	-80	-2.5	224
B	9.9	-93	-4.2	466
C	8.7	-122	-1.4	529
D	12.5	-55	-5.1	389

Unconstrained Market Settlement

In the transmission unconstrained market settlement case the market price is determined simply by matching the total demand to the total generation offers as follows,

$$\begin{aligned}&(10.2\cdot\rho_M - 80) + (9.9\cdot\rho_M - 93) + (8.7\cdot\rho_M - 122) + (12.5\cdot\rho_M - 55)\\ &\quad - (-2.5\,\rho_M + 224) - (-4.2\cdot\rho_M + 466) - (-1.4\cdot\rho_M + 529) - (-5.1\cdot\rho_M + 389) = 0\end{aligned}$$

Solving the above, we obtain the unconstrained market price, ρ_M, and other parameters, as reported in *Table 10.* From *Table 10* we see that the market-clearing price is 35.927$/MWh and this is applicable to all participants irrespective of their location. The generation and demand cleared in each area is shown in columns 3 and 4 respectively. Column-5 shows the net transfer out of an area (if positive) or into an area (if negative).

Table 10. Optimal solution of the unconstrained pool settlement

Areas	Market price, ρ $/MWh	PG, MW	PD, MW	Net Transfer, MW	Transfers, MW			
					A	B	C	D
A	35.927	286.451	134.183	+152.268	-	+52.435	+288.141	-188.308
B		262.673	315.108	-52.435	-52.435	-	0	0
C		190.561	478.703	-288.141	-288.141	0	-	0
D		394.083	205.774	+188.308	+188.308	0	0	-

As seen, area-A and area-D are net exporters while area-B and area-D are net importers. However, note that area-A does import power from area-D, though the net transaction of area-A is positive (denoting a net exporter). The power transaction matrix is shown in the next four columns, showing the power carried by each transmission interconnection[4].

Now let us assume that the transmission links A-B, A-C and A-D, each have a power carrying capacity of 30 MW only, *i.e.* $T_{i,j}^{Max}$ = 30 MW. Then, from *Table 10* we can see that all the transmission links: A-B, A-C and A-D violate the capacity limits. In order to handle this situation, the market operator splits the system into four different price areas as shown in *Figure 13*. The market price is re-evaluated, considering each price area settlement separately now.

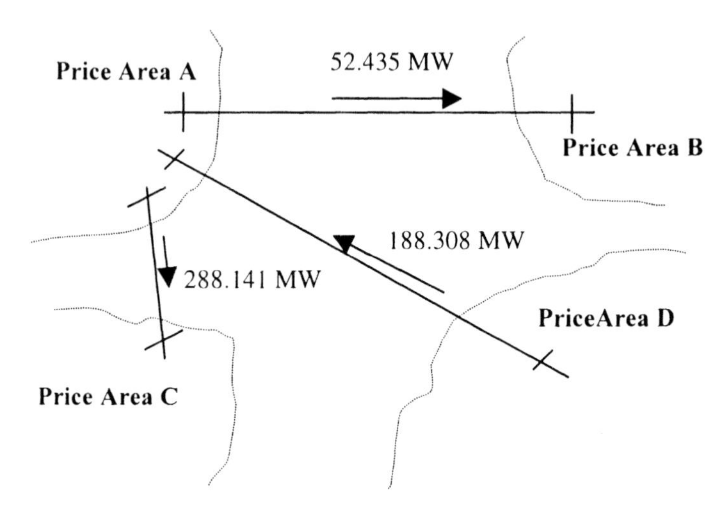

Figure 13. Power transactions in unconstrained market settlement and splitting of the market due to congestion on the transmission links

[4] Note that transmission losses have been neglected

Market Settlement with Congestion Management Constraints

Based on the discussions earlier, the demand-supply balance for each area is resolved separately, and the constraint equations will be of the form,

$$(10.2 \cdot \rho_A - 80) - (-2.5 \cdot \rho_A + 224) - T_{A,B} - T_{AC} + T_{A,D} = 0$$
$$(9.9 \cdot \rho_B - 93) - (-4.2 \cdot \rho_B + 466) + T_{B,A} = 0$$
$$(8.7 \cdot \rho_C - 122) - (-1.4 \cdot \rho_C + 529) + T_{C,A} = 0$$
$$(12.5 \cdot \rho_D - 55) - (-5.1 \cdot \rho_D + 389) - T_{D,A} = 0$$

Note that in the above, $T_{i,j}$ = 30 MW ($\forall$ *i*, *j* $\in$ A, B, C, D) needs to be used so as to impose the transmission constraint. ρ_A, ρ_B, ρ_C and ρ_D are the individual area prices, obtained by treating the demand balance of each area separately, described by the above set equations. *Table 11* shows the optimal market settlement incorporating the transmission congestion constraints.

Table 11. Optimal solution with transmission congestion constraints included

Areas	Area Price, ρ_i, \$/MWh	PG, MW	PD, MW	Net Transfer, MW	Transfer, MW			
					A	B	C	D
A	26.299	188.252	158.252	+30	-	+30	+30	-30
B	37.518	278.430	308.430	-30	-30	-	0	0
C	61.485	412.920	442.920	-30	-30	0	-	0
D	26.932	281.648	251.648	+30	+30	0	0	-

The market settlement process is demonstrated in detail in *Figure 14* for the unconstrained case versus the constrained case of area-A and in *Figure 15* for the unconstrained case versus the constrained case of area-B.

We see from *Figure 14* that the unconstrained system price reduces from \$35.927/MWh to \$26.299/MWh for area-A in the constrained case, since area-A is a generation-surplus area. Thus, the generators receive a lesser payment, thereby discouraging excess generation in area-A while customers benefit from the reduced price, thereby encouraging more demand in this area. This also holds true for area-D, which is also a generation-surplus area in the unconstrained market, and its price is reduced to \$26.932/MWh in the constrained case.

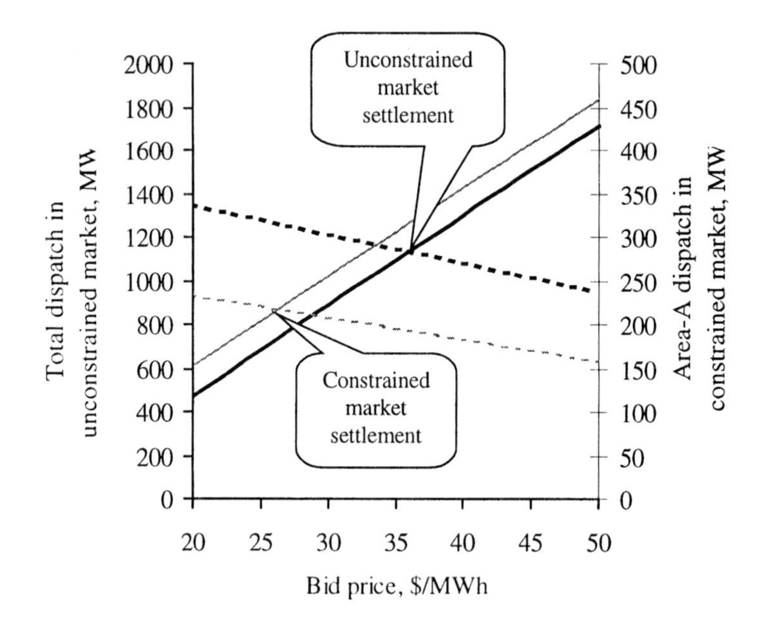

Figure 14. Unconstrained market settlement (shown by the pair of bid-price curves in bold print) vis-à-vis the constrained market settlement in Area-A (shown by the pair of bid curves in lighter print). The constrained market price reduces since area-A is a generation-surplus area. (Note that demand-bid curves, with negative slopes, are shown by dotted lines. The supply-bid curves, with positive slopes, are shown by continuous lines).

On the other hand, *Figure 15* shows that the unconstrained system price increases from $35.927/MWh to $37.518/MWh for area-B in the constrained case, since area-B is a generation-deficit area. Thus, generators receive a higher payment, thereby encouraging more generation in area-B while customers have to pay more, thereby discouraging more demand in this area. This also holds true for area-C, which is generation-deficit in the unconstrained market, and its price is reduced to $26.932/MWh in the constrained case.

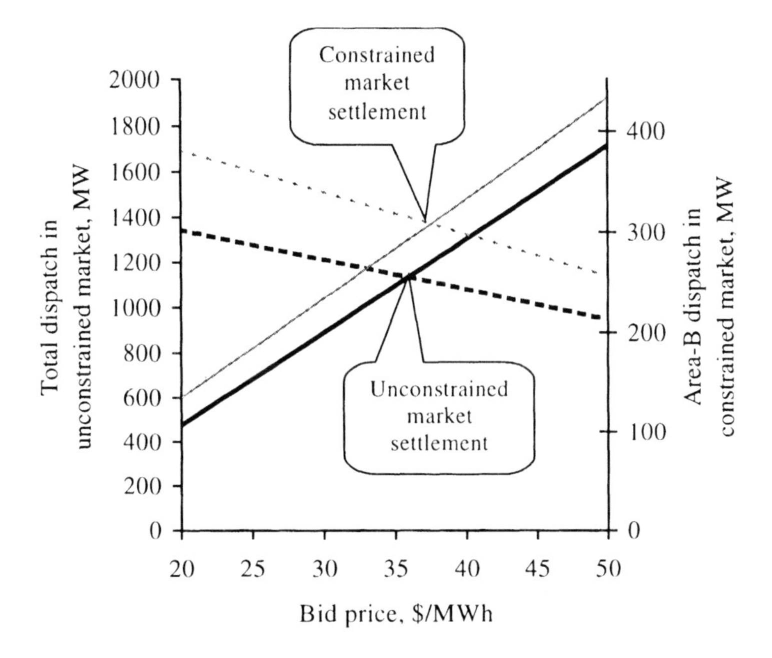

Figure 15. Unconstrained market settlement (shown by the pair of bid-price curves in bold print) vis-à-vis the constrained market settlement in Area-B (shown by the pair of bid curves in lighter print). The constrained market price increases since area-B is a generation-deficit area. (Note that demand-bid curves, with negative slopes, are shown by dotted lines. The supply-bid curves, with positive slopes, are shown by continuous lines).

9.1.2 Counter Trade

Counter trade in simple terms implies a trading in the opposite direction to an existing trade that is causing transmission congestion. In Sweden and Finland, this is the usual way of handling congestion arising from bilateral-market and spot-market trades. Therefore, if there is a bottleneck on the north-south transmission corridor in Sweden due to bulk power transfers from the northern part of the country to the south, the ISO can order to increase the generation levels in the south while reducing the generation in north. These changes in the generation levels are called for by the ISO based on the bids available to it from the balance market[5] and are executed on a real-time basis.

[5] *Balance market* was briefly mentioned in *Chapter-3* and will be further discussed in *Chapter-5*.

There is an additional built in congestion management parameter within the Swedish transmission fee, called the *capacity charge*. This component is based on geographical location. Since bulk of the power flow in Sweden is from north to south, the capacity charge is set higher in the north for generators and in the south for loads. Similarly, the prices are low for loads in the north and generators in the south. The variation is linear, and the generator and load capacity charge curves are reciprocal, *i.e.*, the highest and lowest charges are the same. The range is from 2 SEK/MW for the low (south for generators and north for loads) to 38 SEK/MW for the high (*this was discussed in Section-6.3. Refer to Figure 10 of this chapter*). The variations in charges acts as an economic signal to encourage addition of generation and load to the system in locations that do not increase congestion, thus acting as a long term congestion deterrent.

However, despite the different approaches to congestion management within the Nordic system, both the methods coexist successfully within the same market environment.

10 CONCLUDING REMARKS

In this chapter we attempted to provide an overview of the various issues in transmission transactions in deregulated electricity markets. There have been differing structures of management of the transmission system in the US and in Europe and these have been discussed in the context of Transmission System Operators (TSO) who hold the additional responsibility of an ISO. Various costs incurred by a transmission company to provide access and the rational allocation of these costs to the participating transmission customers are important issues. Most of the debate has been towards the recovery of the total costs in an economically efficient manner. Pricing methods based on embedded costs and marginal costs are discussed. Subsequently, an attempt has been made to cover a few of the country specific cases and the transmission pricing methods unique to those systems. It is interesting to note the longitudinal pricing method being used in Sweden, which implicitly helps in congestion management as well. We also discuss the most recent scheme on international transmission tariffs to be adopted in Europe in the near future.

Security and congestion management are vital issues in restructured power systems and there have been various market-based approaches to handle them. We have discussed some of the ways of handling these aspects. Interruptible load is being actively considered as a security service and it has already been incorporated in the market mechanisms in various countries, as we have discussed. Several other countries are discussing ways of introducing attractive interruptible load contracts, among them Sweden and New Zealand, to name a few. We have also discussed an OPF based method to evaluate optimal incentives for interruptible load contracts. Finally, congestion management in deregulated markets has been examined-particularly, the price area separation method being practiced in Norway, and the counter-trading method being practiced in Sweden, both having proved to be quite successful in managing congestion in these systems.

REFERENCES

1. H. Rudnick and R. Raineri, "Transmission pricing practices in South America", Utilities Policy, Vol.6, 1997, pp.211-218.
2. H. H. Happ, "Cost of wheeling methodologies", IEEE Transactions on Power Systems, Feb.'94, pp.147-156.
3. D. Shirmohammadi, P. R. Gribik, E. T. K. Law, J. H. Malinowski and R. E. O'Donnel, "Evaluation of transmission network capacity use for wheeling transactions", IEEE Transactions on Power Systems, Oct.'89, pp.1405-1413.
4. N. S. Rau, "Certain considerations in the pricing of transmission service", IEEE Transactions on Power Systems, Aug. '89, pp. 1133-1139.
5. A. F. Vojdani, C.F. Imparato, N. K. Saini, B. F. Wollenberg and H. H. Happ, "Transmission access issues", IEEE Transactions on Power systems", Feb.'96, pp.41-51.
6. C. B. Lankford, J. D. McCalley and N. K.Saini, IEEE Task force on Transmission Access and Non-utility Generation, "Bibliography on transmission access issues", IEEE Transactions on Power Systems, Feb.'96, pp.30-40.
7. J. D. McCalley, S. Asgarpoor, T. Gedra, M. Halpin, N. K. Saini and M. H. Schrameyer, IEEE Subcommittee on Open Access Issues, "Second bibliography on transmission access issues", IEEE Transactions on Power Systems, Nov.'97, pp.1654-1659.
8. R. A. Wakefield, J. S. Graves and A. F. Vojdani, "A transmission services costing framework", IEEE Transactions on Power Systems, May '97, pp. 622-628.
9. D. Shirmohammadi, C. Rajagopalan, E. R. Alward and C. L. Thomas, "Cost of transmission transactions: An introduction", IEEE Transactions on Power Systems, Nov.'91, pp. 1546-1556.
10. H. M. Merrill and B. W. Erickson, "Wheeling rates based on marginal-cost theory", IEEE Transactions on Power Systems, Nov. '89, pp. 1445-1451.
11. M. C. Caramanis, N. Roukos and F. C. Schweppe, "WRATES: A tool for evaluating the marginal cost of wheeling", IEEE Transactions on Power Systems, May '89, pp. 594-605.
12. J. S. Clayton, S. R. Erwin and C. A. Gibson, "Interchange costing and wheeling loss evaluation by means of incrementals", IEEE Transactions on Power Systems, Aug. '90, pp. 759-765.
13. Y. Z. Li and A. K. David, "Wheeling rates of reactive power flow under marginal cost pricing", IEEE Transactions on Power Systems, Aug. '94, pp. 1263-1269.
14. Y. Z. Li and A. K. David, "Optimal multi-area wheeling", IEEE Transactions on Power Systems, Feb. '94, pp. 288-294.
15. D. Shirmohammadi, X. V. Filho, B. Gorenstin and M. V. P. Pereira, "Some fundamental, technical concepts about cost based transmission pricing", IEEE Transactions on Power Systems, May '96, pp. 1002-1008.

16. H. G. Stoll, *Least cost electric utility planning*, John Wiley & Sons, 1989.
17. C. W. Yu and A. K. David, "Pricing transmission services in the context of industry deregulation", IEEE Transactions on Power systems, Feb.'97, pp.503-510.
18. R. R. Kovacs and A. L. Leverett, "A load flow based method for calculating embedded, incremental and marginal cost of transmission capacity", IEEE Transactions on Power Systems, Feb. '94, pp. 272-278.
19. M. C. Caramanis, R. E. Bohn and F. C. Schweppe, "The cost of wheeling and optimal wheeling rates", IEEE Transactions on Power Systems, Feb.'86, pp.63-73.
20. I. J. Pérez-Arriaga, F. J. Rubio, J. F. Puerta, J. Arceluz and J. Marín, "Marginal pricing of transmission services: an analysis of cost recovery", IEEE Transactions on Power Systems, feb.'95, pp.546-553.
21. B. L. P. P. Perera, E. D. Farmer and B. J. Cory, "Revenue reconciled optimum pricing of transmission services", IEEE Transactions on Power Systems, Aug. '96, pp. 1419-1426.
22. R. D. Tabors, "Transmission system management and pricing new paradigms and international comparisons", IEEE Transactions on Power Systems, Feb. '94, pp.206-215.
23. R. D. Tabors, "Lessons from the UK and Norway", IEEE Spectrum, Aug.'96, pp.45-49.
24. H. Rudnick, R. Palma and J. E. Fernandez, "Marginal pricing and supplement cost allocation in transmission open access", IEEE Transactions on Power Systems, May '95, pp. 1125-1132.
25. H. Rudnick, M. Soto and R. Palma, "Use of system approaches for transmission open access pricing", International Journal of Electrical Power and Energy Systems, Vol.21, Issue-2, Feb.'99, pp.125-135.
26. J. Perez-Arriaga, H. Rudnick and W. O. Stadlin, "International power system transmission open access experience", IEEE Transactions on Power Systems, Feb. '95, pp. 554-564.
27. Svenska Kraftnät, "The Swedish electricity market reforms and its implications for Svenska Kraftnät", Second Edition, 1997.
28. European Commission, "International exchanges of electricity: Draft rules proposed by the European Transmission System Operator", January 1999.
29. A. I. Cohen, V. Brandwajn and S. K. Chang, "Security constrained unit commitment for open markets", Proceedings of Power Industry Computer Applications Conference (PICA'99), pp.39-44.
30. T. Alvey, D. Goodwin, Xingwang Ma, D. Streiffert and D. Sun, "A security-constrained bid-clearing system for the New Zealand wholesale electricity market", IEEE Transactions on Power Systems, May '98, pp. 340-346.
31. J. W. M. Cheng, F. D. Galiana and D. T. McGillis, "Studies of bilateral contracts with respect to steady-state security in a deregulated environment", IEEE Transactions on Power Systems, Aug. '98, pp. 884-889.
32. J. W. M. Cheng, D. T. McGillis and F. D. Galiana, "Probabilistic security analysis of bilateral transactions in a deregulated environment", IEEE Transactions on Power Systems, Aug.'99, pp.1153-1159.
33. J. Zhong and K. Bhattacharya, "Optimum Var support procurement for maintenance of contracted transactions", Proc. of International Conference on Electric Utility Deregulation and Restructuring and Power Technologies 2000, DRPT2000, London, 2000.
34. H. Wu and H. B. Gooi, "Optimal scheduling of spinning reserve with ramp constraints", Proc. of IEEE Power Engineering Society Winter Meeting, 1999, Vol.2, pp. 785-790.
35. H. B. Gooi, D. P. Mendes, K. R. W. Bell and D. S. Kirschen, "Optimal scheduling of spinning reserve", IEEE Transactions on Power Systems, Nov.'99, pp.1485-1492.

36. T. W. Gedra and P. P. Varaiya, "Markets and pricing for interruptible electric power", IEEE Transactions on Power Systems, Vol.8, Feb.'93, pp.122-128.
37. G. Strbac, E. D. Farmer and B. J. Cory, "Framework for the Incorporation of demand side in a Competitive Electricity Market", IEE Proceedings Generation, Transmission and Distribution, Vol.143, May 1996, pp.232-237.
38. The Electricity Pool of England and Wales, "Demand Side Bidding (DSB-1 Scheme)", Issue 2, 1997.
39. New York Independent System Operator, "Treatment of Distributed Resources: The Accommodation of Price Sensitive Load, Interruptible Load, Dispatchable Load, Distributed Generation and Intermittent Generation in the New York Electricity Market", NYISO-DSM Focus Group, 1998.
40. New York Independent System Operator Ancillary Services Manual, 1999.
41. California Independent System Operator, "Summer 2001 Demand Relief Program", November 2000.
42. Power Pool of Alberta, "Voluntary Load Curtailment Program", November 1998.
43. National Electricity Code Administration, "Code Change Panel: Demand-Side Participation", Consultation Paper, September 2000.
44. K. Bhattacharya, M. H. J Bollen and J. E. Daalder, "Real Time Optimal Interruptible Tariff Mechanism Incorporating Utility-Customer Interactions", IEEE Transactions on Power Systems, Vol. 15, May 2000, pp.700-706.
45. M. C. Caramanis, R. E. Bohn and f. C. Schweppe, "System security control and optimal pricing of electricity", Internation Journal of Electrical Power & Energy Systems, Vol.9, Oct.'87, pp.217-224.
46. R. J. Kaye, F. F. Wu and P. Varaiya, "Pricing for system security", IEEE Transactions on Power Systems, May'95, pp.575-583.
47. R. Rajaraman, J. V. Sarlashkar and F. L. Alvarado, "The effect of demand elasticity on security prices for the poolco and multi-lateral contract models", IEEE Transactions on Power Systems, Aug.'97, pp.1177-1184.
48. R. K. Chaturvedi, K. Bhattacharya and J. Parikh, "Transmission Planning for the Indian Power Grid: A Mixed Integer Programming Approach", International Transactions on Operations Research, Vol. 6, Issue 5, pp. 465-482, 1999.
49. H. Glavitsch and F. Alvarado, "Management of multiple congested conditions in unbundled operation of a power system", IEEE Transactions on Power Systems, Aug.'98, pp.1013-1019.
50. R. S. Fang and A. K. David, "Transmission congestion management in an electricity market", IEEE Transactions on Power Systems, Aug.'99, pp. 877-883.
51. R. S. Fang and A. K. David, "Optimal dispatch under transmission contracts", IEEE Transactions on Power Systems, May'99, pp.732-737.
52. F. D. Galiana and M. Ilic, "A mathematical framework for the analysis and management of power transactions under open access", IEEE Transactions on Power Systems, May'98, pp.681-687.
53. R. D. Christie and I. Wangensteen, "The energy market in Norway and Sweden: Congestion management", IEEE Power Engineering Review, May 1998, pp.61-63.
54. R. D. Christie, B. F. Wollenberg and I. Wangensteen, "Transmission management in the deregulated environment", Proceedings of the IEEE, Feb.2000, pp.170-195.

Chapter 5

Ancillary Services Management

1. WHAT DO WE MEAN BY ANCILLARY SERVICES?

The function of an electric utility is not limited to power generation, transmission and its distribution to customers. It also has to ensure the required degree of quality and safety, undertake preventive measures to ward off contingencies, and perform several other functions. *Ancillary services* are defined as all those activities on the interconnected grid, that are necessary to support the transmission of power while maintaining reliable operation and ensuring the required degree of quality and safety. *Ancillary services* would thus include,

- Regulation of frequency and tie-line power flows
- Voltage and reactive power control
- Ensuring system stability
- Maintenance of generation and transmission reserves

and many others.

According to the North American Electric Reliability Council (NERC) [1, 2], an ancillary service is an *interconnected operation service* that is necessary to effect a transfer of electricity between purchasing and selling entities, and which a transmission provider must include in an open access transmission tariff.

In vertically integrated utility structures where responsibility for generation, transmission and distribution is centralized at one organization, ancillary services are an integral part of the electricity supply and are not separated. However with the deregulation of the power industry, with generation and transmission becoming separate businesses, the system operator often has no direct control over individual power stations and has to purchase ancillary services from ancillary service providers. In such an environment, issues pertaining to pricing mechanisms for such services are extremely important for the proper functioning of the system. There are several operator activities and services, which can come under the purview of ancillary services. The definitions of some services and distinctions between some of them are often unclear.

While the details and definitions of some ancillary services remain vague, the key concepts and purposes of these services are now widely understood and appreciated. However, much work remains in defining a set of services- mutually exclusive and exhaustive, in identifying which services can be provided competitively, which must be under the direct control of the independent system operator (ISO), and which can be obtained from outside the local control area[1].

1.1 General Description of Some Ancillary Services

1.1.1 Frequency Control

Operation of ac interconnections is possible only because a balance is maintained between the supply and the demand within a control area, taking into account any scheduled interchange of power with other control areas. This balance can be maintained by variation of the real power output in generators or through other different procedures.

Most power systems are designed to withstand a loss of generation equivalent to a large generating unit of the system. This indicates that there should be a reserve available to the system that can be activated fast enough, and equivalent to the size of the large unit, to take care of this loss of generation. The concept of area control error (ACE) has been developed as a criterion that indicates the deviation from scheduled interchange plus the frequency bias. Mathematically ACE can be written as,

$$\text{ACE, in MW} = +\Delta P + R \cdot \Delta f = (P_{actual} - P_{scheduled}) + R \cdot \Delta f \quad (1)$$

ΔP = Deviation between the actual and scheduled interchange (+ indicates the direction out of the area)

R = Actual frequency response characteristic of the area in MW/Hz

[1] Control area is an electrical system bounded by an interconnection (tie-line) metering and telemetry. It controls generation directly to maintain its interchange schedule with other control areas and contributes to frequency regulation of the interconnection [1].

Δf = Frequency deviation between the actual frequency and base frequency (50 Hz or 60 Hz).

The first term in the ACE equation denotes the control area's load-generation balance with the rest of the interconnection, and the second term reflects the interconnection's frequency deviation from the reference.

When a disturbance takes place in the system, there is a mismatch in the load-generation balance and the frequency dips. The frequency dip is instantaneously followed by an increase in generation initiated by governor response. This instantaneous generation increase combined with some frequency-dependent load reduction helps arrest any further fall in the frequency. Without any other action, the system will operate at a new steady-state frequency that's slightly less than nominal. This is commonly known as governor regulation action or sometimes, *primary frequency control*.

This will however cause unscheduled flows on the tie line and lead to undesirable ACE accumulation. To restore the system to nominal frequency, the generation set point of the unit should be re-adjusted based on the new generation-load balance. In many instances in the US, this is done through an automatic secondary control action, known as the *Automatic Generation Control (AGC)*, while in most of the European countries this is achieved through manual adjustment of the supply to balance the load. This is thus referred to as *secondary frequency control*.

1.1.2 Reserves Services

Operating reserve in a system is a measure of its ability to increase the generation in order to take care of contingencies such as generation outage or shortfall. The system must be in a position to respond to any such unforeseen event and to ensure that the demand supply balance is maintained. In order to do this, the ISO must arrange for operating reserves in the system, which are capable of providing for additional generation margins with response time ranging from few seconds to several minutes, to bring into application [3].

1.1.3 Reactive Power and Voltage Control Service

System voltages need to be maintained within acceptable limits at all times. In order to achieve this, the ISO has to provide a voltage control service wherein the reactive power sources including generating units can be called upon to provide the requisite support. The amount of reactive power/ voltage control service required is determined from systems-level simulations based on the reactive power support necessary to maintain system voltages within limits.

1.1.4 Black Start Capability Service

A major breakdown of a power system is a contingency that is rare but one that nevertheless does occur. In order to reduce the economical and social consequences it is important to restore power as fast and as securely as possible. A restoration process can be initiated with assistance from neighboring areas by tie lines or at stations with black-start capability.

Two major problems associated with the restoration process are the voltage and frequency control. Voltage control includes the energizing of long transmission lines and the reactive capability available from generators and other reactive sources. Frequency control is associated with system inertia, turbine governors and their control systems and the frequency dependency of the loads.

In a deregulated environment, the problem is further complicated by the presence of multiple independent generators and transmission companies. The ISO, in addition to its other functions, is also entrusted with the responsibility of procuring black-start capability for the system and ensuring that the system is restored back to normal operating state in the shortest possible time. The procurement of black start capability may be done through a competitive process. Technical restrictions on the unit's capabilities and the location of such generators need to be considered while black start capability arrangements are made [4].

1.1.5 Scheduling and Dispatch Services

This service is only applicable to systems in which the ISO performs the scheduling and dispatch functions. The service consists of all administrative tasks required to schedule and control the functions within the boundaries of the ISO control area. Market participants are normally charged a monthly fee for this service. This is not applicable to the bilateral contract-dominated Nordic power systems since generation scheduling is outside the purview of the ISO.

2. ANCILLARY SERVICES MANAGEMENT IN VARIOUS COUNTRIES

The classification and definition of various ancillary services differ across systems and markets. In the following sections we briefly discuss the basic functions of each service in the general sense.

2.1 The US

According to the NERC Operating Policy 10 [1, 2], the various services that are required for safe and secure delivery of power to the customer are the following:

- **For Maintaining Generation and Load Balance**
  - Regulation service
  - Load Following service
  - Contingency Reserve service

- **For Bulk Transmission System Security**
  - Reactive Power Supply from Generation Sources
  - Frequency Response service

- **For Emergency Preparedness**
  - System Black Start Capability

2.1.1 Generation and Load Balancing

2.1.1.1 Regulation and Load Following Services

The basic objective of the ISO is to maintain a minute-to-minute balance between the generation and load. To this effect, two services, *regulation* and *load following* have been specified. Regulation is the minute-to-minute adaptation of a generator output to meet the imbalance between total supply and demand in the system. This instantaneous response of a generating unit is usually achievable through the use of the governor-droop characteristic or automatic control signals from the control area determining the required change (up and down) to the real power output to correct the Area Control Error to within bounds. Load following on the other hand, involves balancing the generation-demand imbalance at the end of a scheduling interval. This may be automatic or manual and basically track a somewhat long-term variation of load (*Figure 1*).

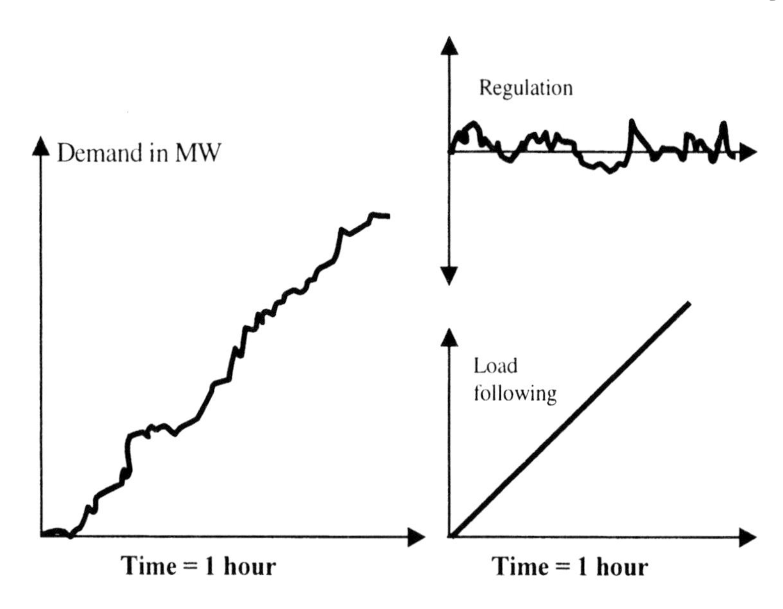

Figure 1. Regulation is the minute-to-minute change in generator output to balance the demand while load following is the adjustment in generation at the end of scheduling interval.

2.1.1.2 Contingency Reserve Services

These services are activated in times of contingencies and are meant to restore the generation and demand balance in the system. These are classified into two categories: (a) spinning reserve services and (b) supplemental reserve services, as described below.

Spinning reserve is the reserve available from a synchronous generator on-line and generating at a level lower than its maximum rated capacity. This reserve is available at a short notice. However, the ISO needs to ensure that spinning reserves are spread out over the system uniformly so that the reserves are not affected by transmission constraints. Ideally these reserves should be available instantaneously and within 10 minutes.

Supplemental reserves include those generating capacities that are available within about 10 minutes after the disturbance. This includes generators with fast start-ups such as gas or oil-fired generators and hydro units. Customers can also participate in providing for supplementary reserves by way of contracting for interruptible loads.

2.1.2 Bulk Transmission System Security

2.1.2.1 Reactive Power Supply from Generation Sources

Reactive power support is essential for the maintenance of a uniform voltage profile on the system during normal and emergency conditions. This is accomplished by coordinating the various reactive power resources available to the operator. Among others, load power factor correction, reactive power compensation through capacitors, reactors, *etc.*, reactive power support from generators are some accepted methods for voltage control.

NERC's proposed standards on interconnected operations services have defined that reactive power from generating sources only qualify as ancillary services and hence for financial compensation from the ISO. Voltage control services provided by other system components are not considered ancillary services.

2.1.2.2 Frequency Response

NERC has argued that this service would come under the purview of bulk transmission system security and should be deployed in response to a *reliability* need of the interconnection. Frequency response is the immediate governor response resulting from a change in interconnection frequency. When there is a disturbance in the system and the frequency declines; it is the responsibility of each control area to make available certain amount of generation to respond to the frequency deviation instantaneously, based on governor-droop characteristics. Although NERC does not specify a minimum limit on the frequency response characteristic for each area, each area is expected to maintain some frequency response capability.

2.1.3 Emergency Preparedness

2.1.3.1 System Black Start Capability

NERC has defined this service as an ancillary service and hence it is the responsibility of the ISO to arrange, provide for and deploy system black start capability as and when required to energize the transmission system following a blackout.

Research on electric power ancillary services by Hirst and Kirby [5, 6] outline the US scenario with regard to defining the various terms and explaining their intricate differences. A detailed survey on the cost of ancillary services across 12 major electric utilities in US was carried out that revealed that on the average, the total cost of ancillary services was about

0.41 cents/kWh, which is 10% of the generation and transmission costs [6] (*Figure 2*).

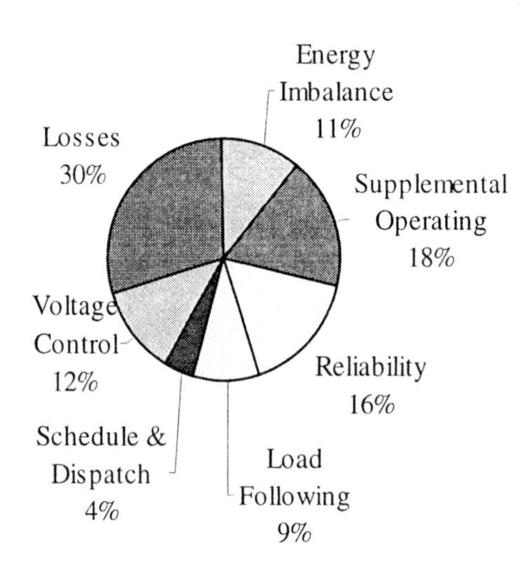

Figure 2. Break up of costs incurred in providing for ancillary services in 12 utilities in US. The figures are in percentage of total ancillary cost assumed 100% [6]

2.2 UK

The National Grid Company (NGC) carries out the role of the ISO in the UK power pool and hence is also responsible for making arrangements with regards to ancillary services. NGC maintains the transmission system security by devising appropriate dispatch schedules with the help of generators' bid data. However, since the ISO can only schedule offered plants, it needs to ensure that sufficient capacity is always available both at the systems level and locally. This requires the procurement and management of various ancillary services. NGC contracts for ancillary services to enable voltage and frequency control standards to be maintained, as well as other services such as the black start capability. The payment for these services is recovered by NGC through the pool.

Gencos, regional electricity companies, large consumers or external pool members can supply these services. One of the key tasks of the *ancillary services business* is to encourage competition in the provision of ancillary services. The Grid Code details the technical operational requirements of

NGC and defines two major categories of ancillary services. These are briefly described below [7].

2.2.1 System Ancillary Services

System ancillary services are fundamental to the satisfactory operation of the system and all centrally dispatched generating units connected to either the NGC transmission system or a supplier's distribution system in England and Wales must be able to provide these services. System ancillary services are classified in two categories as given in *Table 1*.

Table 1. System ancillary services: classifications

Part-1: Those services, which all generators are obliged to provide.	**Part-II: Services that need not be provided at every site. NGC buys these services from sites based on specific contracts.**
➢ Reactive energy Other than that supplied by means of synchronous or static voltage compensators	➢ Operating margin (for frequency control) *e.g.* the capability of a gas turbine to fast start
➢ Operating margin (for frequency control) Capability to provide additional output from a generating unit at short notice	➢ Black start capabilities

2.2.2 Commercial Ancillary Services

The second category of ancillary service comprises the commercial ancillary services. The genco is not obliged to provide these services, which are, hence, subject to commercial agreements. Such services are as under:

- ➢ Reactive energy: supplied by synchronous or static voltage compensators
- ➢ Operating margin: provided by pumped storage units, load reduction, stand-by generation, etc.

2.3 Australia

In Australia the National Electricity Market Management Company (NEMMCO) is responsible for the overall transmission system security and hence for procurement and scheduling of various ancillary services. Seven services are recognized for procurement and hence for financial compensation by NEMMCO and these are categorized in five system control areas as follows [8]:

- ➢ Frequency control services
  - *Automatic generation control:* This service is required from a generator within five minutes of the ISO request.

- *Governor control:* This service is based on the generator's governor droop characteristic and aids in frequency control, acting in a one-second to one-minute time frame.

- *Rapid generator unit loading:* This service requires generators not in operation to synchronize with the system within a five-minute time frame.

- *Load shedding:* This service requires curtailment of load following extreme frequency variations. The loads are expected to automatically respond and disconnect within a one-minute time frame.

➢ Voltage control services
  - *Reactive power:* This service is provided to keep system voltages at desired levels by injection or absorption of reactive power at different nodes in the system. Only the reactive power provided by generators and synchronous compensators is treated as ancillary service and is eligible for payment.

➢ Stability control services
  - *Rapid generator unit unloading:* This requires a generator to unload rapidly from its normal operating condition in order to preserve system stability. This may be required under certain circumstances and is recognized as a system stability service.

➢ Network loading control services
  - *Automatic generation control*
  - *Load shedding*

  These have been discussed above. These two services are also recognized as network loading control services.

➢ System restart services
  - *System restart:* This is similar to the black start capability service discussed earlier. This requires that the generator is able to supply the transmission system following a complete system failure.

2.3.1 Payment Mechanisms

The payments to the ancillary service providers have been organized in the Australian market based upon the type of service provided and the service providers' costs, including opportunity costs if any. Four categories of payment mechanisms are recognized and one particular service may call for any one or more of the components for compensation.

- Availability payment: This is made to those services that require the concerned provider's preparedness for providing the services when called for. Such payment is allowed for system restart services and reactive power services from generating units.

- Enabling payment: This payment is made if a concerned service has been enabled (or activated) by the ISO for use. This applies to all the ancillary services except system restart and reactive power from generators. Reactive power service from synchronous compensators is eligible for this payment though.

- Usage payment: This payment is based on actual use of the concerned ancillary service by the ISO and applies only to the rapid loading and rapid unloading services provided by generating units.

- Compensation payment: This payment is based on the provider's opportunity costs and is paid when the concerned provider has been constrained from operating according to its market decisions.
  This payment is applicable to all categories of services except the reactive support service from synchronous compensators and system restart services.

Accordingly, the cost incurred by the ISO in providing for such services are shown in *Figure 3*. As seen from the figure, system frequency control compensation and reactive power support services accounted for the highest share in the total ancillary services costs, together totaling for more than 50%. AGC and governor action services also accounted for substantial portions, together totaling for more than 28% of the total ancillary service costs.

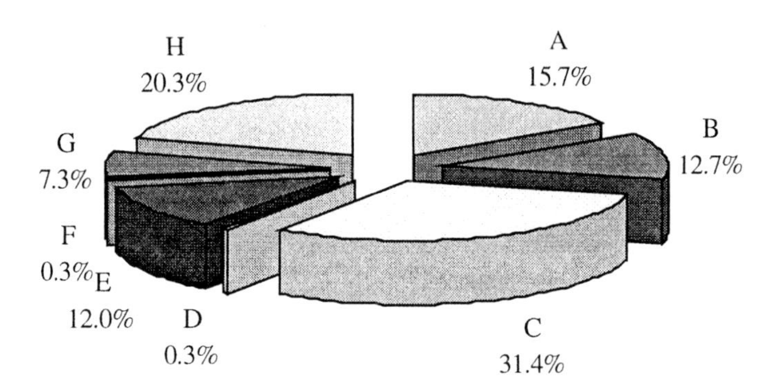

Figure 3. Ancillary service costs in NEMMCO in 1998-99 Explanation of Legends: A: AGC services, B: Governor action, C: Frequency control compensation, D: Load shedding compensation, E: Rapid generator unit loading, F: Rapid generator unit unloading, G: System restart, H: Reactive power. Data source: [9]

2.4 Sweden

In Sweden the ISO, Svenska Kraftnät, manages both the reserve and regulation services through the regulating markets. It additionally pays balancing entities for governor response and load matching on fixed contracts. Presently, no compensation is paid to generators for reactive power. Generation tripping is compensated for when it is ordered in Norway. Load shedding is recognized as an ancillary service, but is not compensated for. No other ancillary service is recognized [10].

In order to maintain the demand-supply balance on a minute-by-minute basis, a two-tier approach is adopted. The primary frequency control is based on automatic frequency control on generators in suitable power stations. Svenska Kraftnät purchases long-term contracts from generating companies for such primary regulation.

A *balance service* is used for secondary regulation of frequency. This is done by taking in regulating bids (volume and price) from generators willing to quickly[2] increase or decrease generation, or even consumers willing to increase or decrease consumption. The bids for regulation are arranged in price order to form a "staircase" for each operating hour. When regulation is

[2] Within about 10 minutes

needed the system operator activates the most favorable bid for regulating up or down (*Figure 4*) [10].

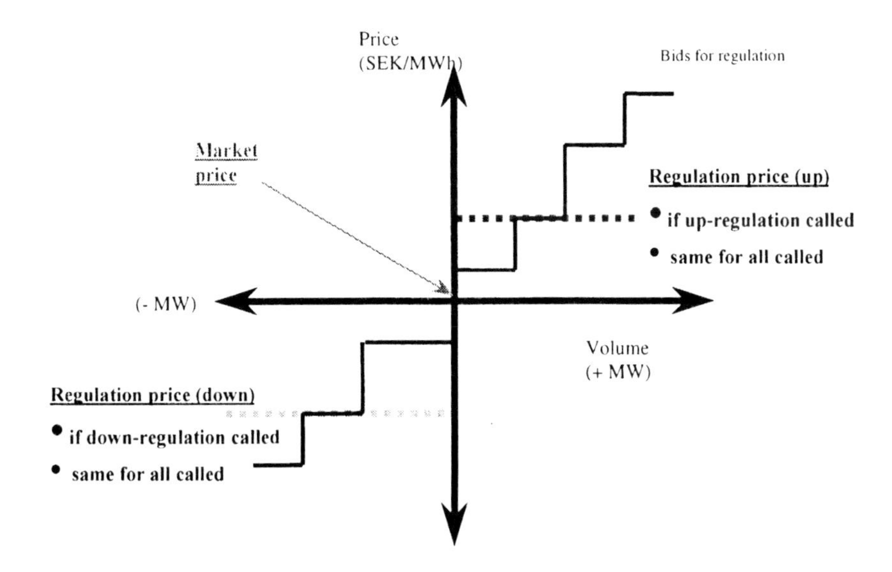

Figure 4. The balance market bids available to the Swedish ISO

If the frequency is lower than nominal, up-regulation bids are activated and the ISO buys the up-regulation service in steps, and the highest activated up-regulation bid becomes the uniform up-regulation price for all providers of upward regulation service. If frequency is higher than nominal, down regulation is desired and now the ISO sells down-regulation service to those bidders, in steps, and the lowest activated down-regulation bid price becomes the uniform down-regulation price for all down regulation service providers (*Figure 5*). At the end of the scheduling interval the net energy balance of each entity is calculated and financial settlements carried out.

This method for secondary regulation is combined with primary regulation based on automatic frequency control on generators in suitable power stations.

At the end of each hour, the costs are distributed based on the individual players' unbalance data and the type of regulation (up or down regulation) called for by the system operator. The hourly regulating price is fixed as the price for the most expensive measure (regulating up- balance service purchases) or least expensive measures (regulating down- balance service sells) utilized during the hour.

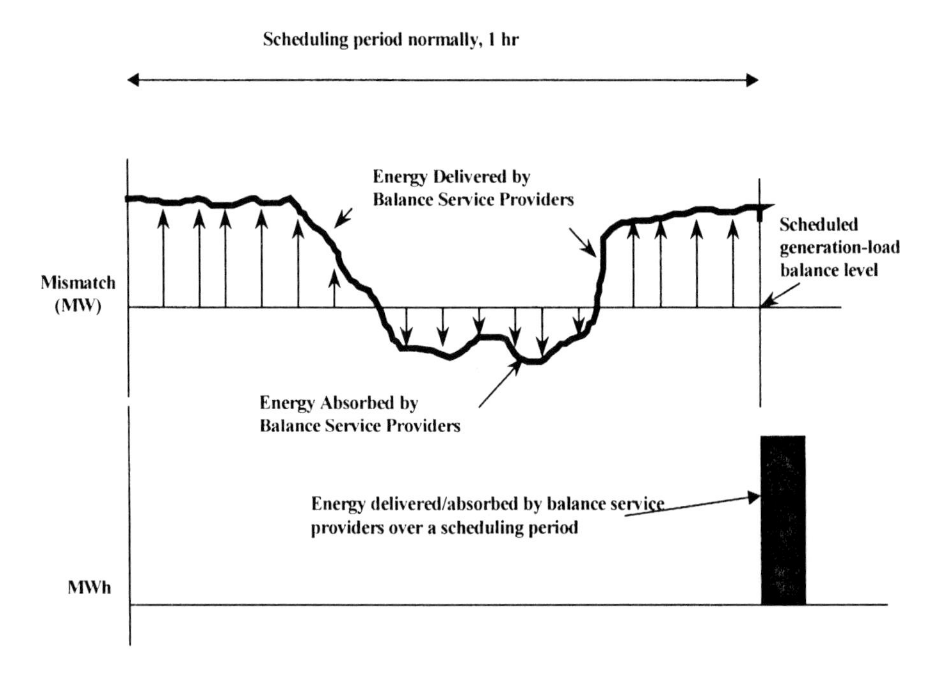

Figure 5. Energy balance by the ISO

2.4.1 Conventions and Practices for Frequency Control in the Nordel[3] System

The nominal frequency in the Nordic system is 50Hz and it is stipulated that fluctuations during normal system operations should not exceed ±0.1Hz. The nuclear units are set to trip below a frequency of 47.5 Hz, hence the frequency should, under no circumstances, go below this. This situation could be alleviated by load-curtailment schemes. At frequencies higher than a certain point above 50 Hz, the thermal units are set to reduce their outputs automatically.

The primary frequency control has the responsibility to cover imbalances due to spontaneous load variations and mismatches between planned generation and actual load level. The total primary frequency control reserves have to meet the following requirements for the whole Nordel system:

- Total reserve margin at 50 Hz: 600 MW
- Total gain, at least: 6000 MW/Hz

[3] Refers to Norway, Sweden, Finland, Denmark and Iceland.

Each country in the Nordel system is responsible for a part of the margin and gain in relation to the annual energy consumption of the system. In the context of US it is worth mentioning that there is no requirement set out by NERC regarding a minimum primary regulation capacity to be available for each system.

Technically, the capacity margin required for primary frequency control, is provided by hydro units. The various operator actions that are initiated when frequency in the Nordel system deviates from its nominal value are shown in *Figure 6* [11].

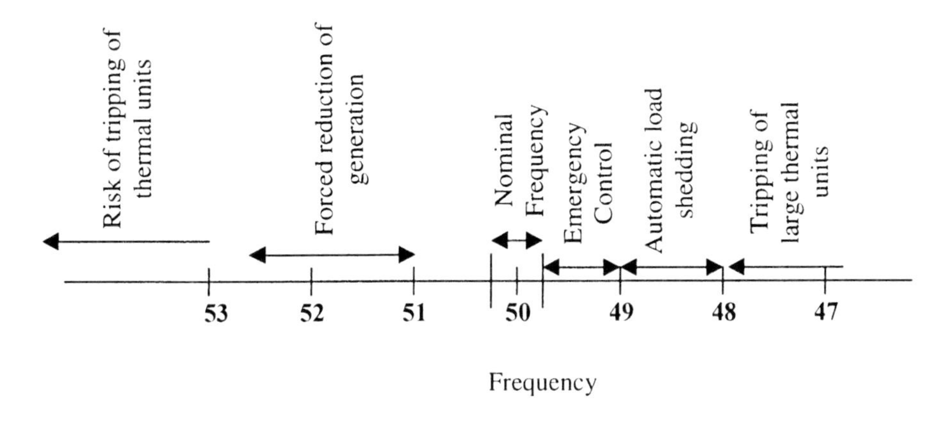

Figure 6. Frequency range and operator action in the Nordel system

2.5 Check List of Ancillary Services Recognized by Various Markets

In the previous sections we have seen that ancillary services have been defined, treated, and paid for in different ways across the deregulated electricity markets over the world. While there are many dissimilarities in the way these are handled, there are some commonalties too, and we attempt to summarize those features in *Table 2*.

Table 2. Ancillary services and how they are handled in various markets

Service	NGC	Svenska Kraftnät	NEMMCO	NYISO
Scheduling and dispatch, system control	Recognized and provided	Scheduling and dispatch not recognized. System control service is through bilateral arrangements with network companies	Not recognized	Recognized. Provided by NYISO. Pricing based on embedded costs.
Voltage support service	Recognized and provided through the Grid Code.	Not provided by ISO. No payment scheme. Certain operating rules apply to network companies. Discussed in detail later.	Recognized and paid to generators and synchronous compensators only	Recognized and priced based on embedded costs to generators only.
Regulation and frequency response service	Provided through the balance service	Primary regulation through bilateral contracts. Secondary regulation through the balance service	Recognized and compensated	Recognized and priced based on the market
Operating reserve service		Within the purview of the balance service	Recognized and compensated	Recognized and price based on the market
Black start capability service	Provided	Provided through bilateral contract.	Recognized and compensated	Recognized and price based on embedded costs

3. REACTIVE POWER AS AN ANCILLARY SERVICE

In this section we examine how reactive power support over the system is managed in various deregulated electricity markets. This varies for each system in the way contracts are framed, or payment mechanisms are devised or the market is operated.

3.1 Reactive Power Management in Some Deregulated Electricity Markets: A Review

3.1.1 UK

3.1.1.1 Reactive Payment Arrangement

In the UK electricity market National Grid Company (NGC) handles the role of the ISO. The *Grid Code* places a minimum obligation on all generating units, with a power generating capacity more than 50 MW, to provide a basic (*mandatory)* reactive power service. In order to receive payment for this service, the generators must enter into a *Default Payment Mechanism (DPM)*. Alternately, the generators can offer the mandatory reactive power service through the tender market by structuring their bids to reflect the value that they perceive their service is worth. This way of meeting the mandatory Grid Code with a market mechanism is termed as *Obligatory Reactive Power Service (ORPS)*. The income a generator could receive by providing reactive power varies according to the relative need and the number of generators that can provide the service within a zone [12, 13].

Further, generators with reactive power capability in excess of the Grid Code can offer an *Enhanced Reactive Power Service (ERPS)*.

The DPM was initially (in 1997/98 when the scheme was started) based on two components- a capability payment component and an actual utilization- based payment component with a ratio of 80:20. This ratio underwent a staircase phasing of the capability component and since April 2000, the ratio has been 0:100 and DPM is based on metered reactive utilization only.

Under the new arrangements, a reactive power market has been formalized by NGC by inviting tenders, which can be for either ORPS or ERPS. Any prospective service provider, irrespective of whether it receives payments under the DPM arrangement or not, can offer a tender. In this way the provider has greater assurance of a given level of income. Unlike the DPM, the bidders are able to offer specific prices for capability and utilization, thereby allowing greater flexibility to offer payment terms that are more cost-reflective of the actual service provided. There are two tender processes a year, which start on 1 April and 1 October, respectively.

3.1.1.2 Structure of Bid Offers in the Tender Market

The tender bids submitted by reactive power service providers comprise two components (a) capability price component (*Figure 7*) and b) utilization price component (*Figure 8*).

When a reactive service provider bids for capability prices, it may choose to bid prices for both leading and lagging MVAr capability or just one of them. It may also bid for two types of capability price bids (a) *synchronous capability price* and (b) *available capability price*. *Figure 7* gives an example cost function that could be offered for synchronized and available capability. For each type of capability price, generators can offer up to three incremental prices for both leading MVAr capability and lagging MVAr capability.

For the utilization bid price, the criteria are similar, *viz.*, up to three incremental prices can be offered for leading and lagging reactive power. Cost function for reactive power utilization is shown in *Figure 8*.

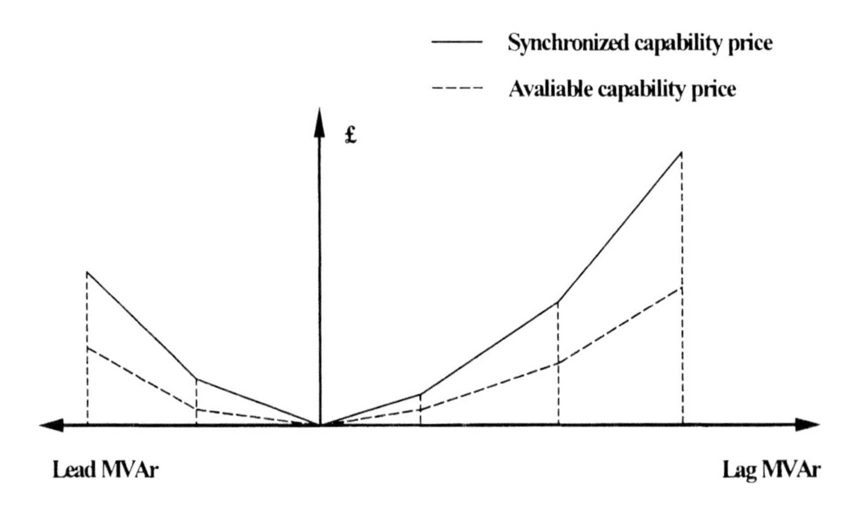

Figure 7. Tender bid price structure for synchronized and available reactive power capacity

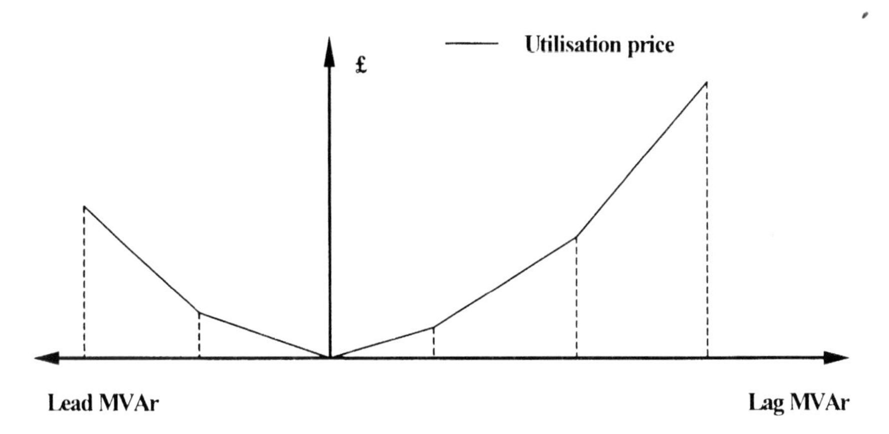

Figure 8. Tender bid price structure for reactive power utilization

Some evaluations have been given in [14] and [15] to describe the bidding trends in the reactive power tender markets for both ORPS and ERPS covering the periods April to September 1999 and October to March 2000 respectively. From the analysis the following observations are made:

- As of April 2000, there were a total of 95 generating units with a reactive power agreement with NGC within the UK reactive power tender market system.

- All the tenders were for the provision of ORPS and none for ERPS. It would be desirable for NGC to receive tenders for ERPS also.

- It was apparent that at least 50% of the bidders were seeking greater remuneration of their reactive power capability although still broadly following the payment profile of DPM.

- A few tenders selected a simple structure of available capability payment, reflecting a linear payment rate per MVAr of reactive power available.

- Some offered a 'flat' available capability payment rate. This type of tenders gave little incentive to NGC to select these generators to maintain capability.

- Majority of the tenders selected the payment structure based on steeper incremental capability prices at higher MVAr outputs. These tenders give clear signal of generators' preferred operating range.

- In a number of the marginal cases, the steep incremental capability prices may influence NGC to decide whether to accept or reject a market agreement.

- Most tender offers included capability prices for hours available, while a significant number of tenders were seeking payment on the basis of the hours synchronized.

3.1.2 New York ISO

In the New York power system, the New York ISO (NYISO) is responsible for providing reactive power support service, and the service is provided at embedded cost-based prices. Generating resources, which operate within their capability limits, are directed by the ISO to produce or absorb reactive power to maintain system voltages within limits. The pricing method for the reactive support service is an embedded cost based price [16].

The transmission customers get reactive support service from the NYISO and pay it for the reactive power service associated with energy withdrawals from the transmission system. The NYISO computes the reactive power support service rated by summing all its payments to suppliers that provide reactive power support. It includes:

- Total annual embedded costs for payment
- Any applicable lost opportunity costs to provide reactive support service
- Total of prior year payments to suppliers of reactive power service less the total of payments received by the NYISO from transmission customers in the prior year for reactive power service.

This sum is divided by the annual forecasted transmission usage for the year as projected by the NYISO including the system load, exports and power wheelings. NYISO calculates this payment hourly. Transmission customers engaged in power wheeling or exporting, pay the NYISO a charge, which is equal to the rate determined as mentioned above, multiplied by the wheeled energy at that hour.

3.1.2.1 Payment for Reactive Power Service

Payments to generators and synchronous condensers for reactive support service are based on the following rules:

- Annual fixed charge rate associated with resource capital investment

- Current capital investment of the resource allocated for supplying reactive support service

- For generators: based on tested power factor, current capital investment of generator's turbine-generator equipment, current capital investment or generator's accessory electrical equipment

- For synchronous condensers: based on current capital investment of synchronous condenser equipment

➢ Operating and maintenance expenses for supervision and engineering allocated for reactive support service
  - For generators: based on operating and maintenance expenses multiplied by 30%*(1-power factor)

  - For synchronous condensers: based on operating and maintenance expenses

The above embedded cost calculation methodology is used to calculate payment to all reactive power providers. The NYISO calculates and makes payments monthly. Suppliers meeting the requirements to supply installed capacity, receive $\frac{1}{12}$th the annual embedded cost payment. Suppliers whose generators are not under contract to supply installed capacity and suppliers with synchronous condensers receive $\frac{1}{12}$th the annual embedded cost payment rated by the number of hours that generators or synchronous condensers operated in that month.

Non-utility generators (NUG) operating with energy or capacity contracts, receive payments for their reactive power support services as discussed below.

➢ If the NUG *provides installed capacity*, its payment is given by:

$$PC = \frac{1}{12}(AR) \times \min\{Q_{Contract}, Q_{Capability}\} \quad (2)$$

➢ If the NUG *does not provide installed capacity*, the payment is as follows:

$$PC1 = PC \times \frac{N_{Operating}}{N_{Total}} \quad (3)$$

AR average annual rate for NYISO payment to voltage support service, \$/MVAr

$Q_{Contract}$ contracted reactive capability, MVAr
$Q_{Capability}$ tested reactive power capability, MVAr
$N_{Operating}$ number of hours in a month NUG providing the reactive support service
N_{Total} number of hours in the month

3.1.2.2 Payment for Lost Opportunity Cost

If the NYISO dispatches or directs a generator to reduce its real power output in order to allow the unit to produce or absorb more reactive power, the generator may receive a component of payment accounting for the Lost Opportunity Cost (LOC). The method for calculating the LOC is based on the following factors:

- Real Time Long-term Based Marginal Price (LBMP)
- Original dispatch point
- New dispatch point
- Bid curve for generation supplying reactive power service

Figure 9 describes the calculation of the LOC for a generator, which decreases its real power output to provide more reactive power service.

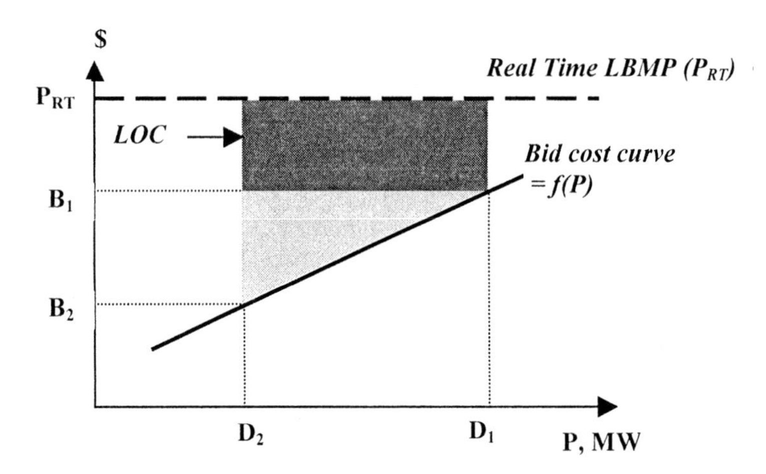

Figure 9. Method for calculating lost opportunity costs

P_{RT} Real Time Long-term Based Marginal Price (LBMP)
f(P) Bid curve for generation supplying voltage support services
D_1 Original Dispatch Point
D_2 New Dispatch Point with increased reactive power output
B_1 Bid price at original dispatch point
B_2 Bid price at new dispatch point

As the real power output is decreased, the generator receives lesser revenue from the sell of energy although, by way of this reduced generation, it saves some generating cost. The reduced income for the generator can be described by (4).

$$\text{Reduced Income for Generator} = \text{Revenue loss - Savings from reduced generating cost} \quad (4)$$
$$= P_{RT}(D_1 - D_2) - \int_{D_2}^{D_1} f(P) \cdot dP$$

Note that, the reduced income for the generator also equals the savings to the ISO. The savings for the generator from reduced real power output is given by (5),

$$\text{Saving for Generator} = B_1(D_1 - D_2) - \int_{D_2}^{D_1} f(P) \cdot dP \quad (5)$$

The LOC of the generator equals the difference between (4) and (5).

$$LOC = (P_{RT} - B_1) \times (D_1 - D_2) \quad (6)$$

3.1.3 Australia

The Australian electricity market and its ISO, the National Electricity Market Management Company (NEMCO) recognizes only that reactive power provided by synchronous generators and synchronous condensors, as ancillary services and financial compensation is made available to them for their service provisions [17].

All reactive power ancillary service providers are eligible for the *availability payment component-* for their preparedness in providing the service when called for. Further, the synchronous compensators also receive an *enabling payment component-* when their service is activated by the ISO for use. On the other hand, a synchronous generator receives the *compensation payment component-* based on its opportunity cost, and gets paid when it has been constrained from operating according to its market decisions. The total payment for reactive power service is shown in *Figure 10*.

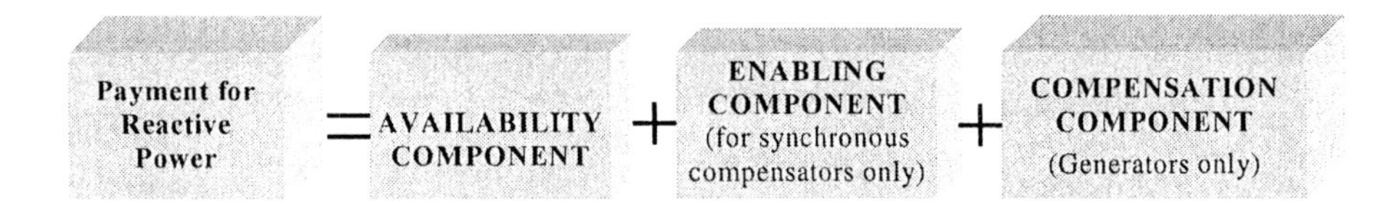

Figure 10. Components of payment for reactive power

3.1.3.1 Mandatory and Ancillary Reactive Power Services

The provision for reactive power from generators is separated into two categories:

- The mandatory reactive power support, and
- Reactive power as an ancillary service

As explained in *Figure 11*, it is *mandatory* for the generators to provide reactive power within the operating power factors of 0.9 lagging and 0.93 leading. Beyond this mandatory component, is the ancillary service component, which the generators offer as per their discretion. However there is a portion beyond the *ancillary service component,* which is left undefined [18].

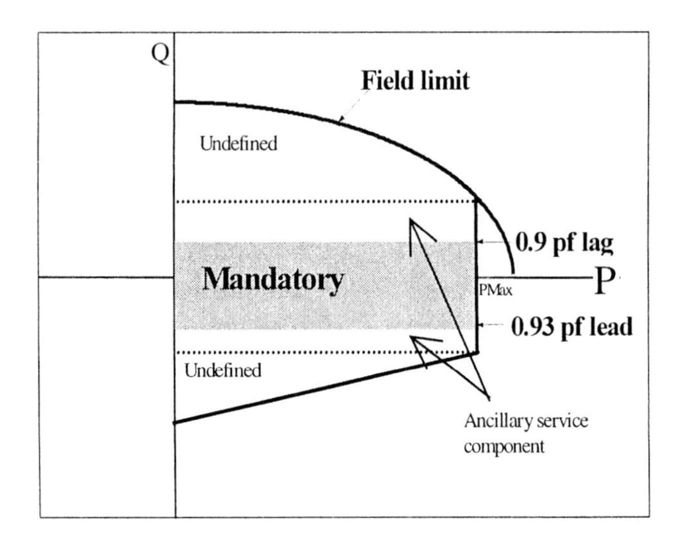

Figure 11. Generator reactive power definitions in the Australian market

The basic voltage control scheme adopted in NEMMCO is as follows [19]:

- Various energy management system (EMS) functions (such as load flow analysis) are used to determine the reactive power requirement in the system

- Reactive power support elements such as capacitor banks, reactors and SVC are switched as required

- Next, reactive power support from generators that are currently on-line, is used to the extent that their normal output is not restrained. Here, those generators not contracted for ancillary services are also called in to provide the *mandatory* amount of reactive power. Those contracted for reactive power can be asked for an amount more than what is mandatory, subject to financial compensation.

- Next, synchronous compensators for the specific area from a merit order based on enabling prices

- If further reactive power is required, the units' real power generation might be constrained

If the total reactive support available from all sources as discussed above is insufficient to ensure system security under certain conditions, market trades can be curtailed.

3.1.4 Finland

The Electricity Market Act and the point of access tariff in 1995 opened the Finish electricity market for competition and in June 1998, Finland became a price area on the NordPool exchange. The Finnish ISO, Fingrid is responsible for maintenance of system voltages at technically and economically optimal levels during both normal and disturbance conditions. Accordingly, it supplies reactive power as per the general supply principles concerning reactive power. The voltage level of the main grid is controlled using reactors and capacitors. The voltage ratio between different voltage steps is controlled with tap changers of the transformers.

3.1.4.1 Reactive Power Reserve Service

Fingrid is also responsible for the maintenance of adequate reactive power reserves in the Finnish power system. This is done through the use of its own resources and also by acquiring reactive reserves from independent parties [20]. This provision for reactive power reserves is a *mandatory* service as of now. It is likely that a tariff mechanism for financial compensation will be in place for this service by the year 2002.

As per the guidelines, generators of more than 10 MVA rating, are required to maintain reactive power reserves during the normal status of the power system:

- For generators connected to 220 kV and 110 kV grid, the *mandatory,* momentary reactive power reserve should not be less than half of the calculated reactive capacity corresponding to a power factor of 0.9. The rest can be used as a commercial service.

- For generators connected to the 400 kV grid, the entire reactive capacity is required to be placed under momentary reserves and *mandatory*, with the exception of that amount consumed by transformers and the plant itself.

- For generators connected to the grid at voltage levels less than 110 kV, half of the reactive power intake capacity at the generator's voltage level- is also required to be reserved as momentary disturbance reserve and *mandatory*.

3.1.5 Sweden

Reactive power service is provided on a *mandatory* basis in Sweden and as of now, there is no scheme for financial compensation to the providers of this service.

The reactive power exchange on the **national grid** is controlled by instructions from Svenska Kraftnät, the Swedish ISO. It is recommended that reactive power flow between different parts of the grid be kept near zero. The ISO has the right to the supply of reactive power from spinning generators connected directly to the national grid.

The Swedish electricity system is characterized by bulk power flows from the north, where a major share of generation is located, to the south, where most of the load centers are, over long distance transmission lines. As reactive power can not be transmitted over such distances, it should be provided by local sources. Svenska Kraftnät owns the national grid (400 kV and 220 kV), while the regional and local network companies operate the sub-transmission and distribution network (130 kV and less).

The regional network companies are responsible for voltage control in their respective areas. Under normal conditions the regional network operators use as much static reactive power production as possible. Large generators are rarely used for secondary voltage control and are reserved for serious situations. Such units operate at a constant reactive power output, with a stable operating point considering vibration and losses.

3.1.5.1 Formal Agreements for Reactive Power Transfer Over the Grid

For power transactions over the network, Svenska Kraftnät enters into formal agreements for reactive power exchange with independent generators and regional networks. Agreements for feeding power into the national grid are mostly undertaken with producers and in certain cases also with regional networks. Following are some of the standard set of agreements:

- A hydro unit connected directly to the national grid is required (*mandatory)* to be able to inject as well as absorb reactive power as per the following limits.

$$\text{Reactive Injection} = \left(\frac{1}{3}P_{Max}\right)_{HV} \text{ to } \left(\frac{1}{3}P_{Max}\right)_{LV} \quad (7)$$

$$\text{Reactive Absorption} = \left(\frac{1}{6}P_{Max}\right)_{HV} \text{ to } \left(\frac{1}{6}P_{Max}\right)_{LV} \quad (8)$$

- A thermal unit connected directly to the national grid is required (*mandatory)* to maintain capability of reactive power injection as per the limits. However, it has no requirement on absorption of reactive power.

$$\text{Reactive Injection} = \left(\frac{1}{3}P_{Max}\right)_{HV} \text{ to } \left(\frac{1}{3}P_{Max}\right)_{LV} \quad (9)$$

$$\text{Reactive Absorption} = (0)_{HV} \quad (10)$$

- A regional network with agreement to inject real power into the national grid, is required to maintain a capability to inject reactive power, depending on the instantaneous real power injection, as given below.

$$\text{Reactive Injection} = \left(\frac{1}{3}P_{instantaneous}\right)_{HV} \quad (11)$$

There is no requirement on absorption of reactive power from the national grid.

$$\text{Reactive Absorption} = (0)_{HV} \tag{12}$$

Also there is no specific requirement from a generator connected to the regional grid.

- In the case of a regional network with agreement for drawing real power from the national grid, there is no requirement for injection or absorption of reactive power to/from the national grid.

$$\text{Reactive Injection} = (0)_{HV} \tag{13}$$

$$\text{Reactive Absorption} = (0)_{HV} \tag{14}$$

3.2 Defining Scope of the Service is Important

Reactive power support services can be used by the ISO from several quarters during a day's operation and it is thus very important to clearly classify those that fall under the category of ancillary service, and those that don't. As we know, reactive power support is essential for voltage control on the system. Voltage control can be split into the same set of services as real power because of the symmetry between real power and reactive power. The first control service, termed as *primary control*, acts to compensate rapid fluctuations occurring on a time scale of up to a few minutes while the second hierarchical level is the *secondary control* where slow variations in voltage on a time scale of up to several hours is compensated.

- **Primary Voltage Control Services (automatic)**
  - Generators
  - Synchronous condensors
  - SVC, Tap changers, etc

- **Secondary Voltage Control Services**
  - Generators
  - Synchronous condensors
  - Shunt reactors / capacitors
  - Power lines, Tap changers, series capacitors

We have seen in the discussions on the UK, Australian and the New York markets that, reactive power from synchronous generators and synchronous condensors qualify as ancillary services. The NERC has also applied the same convention. However, there is an indication that some of the markets in Europe will have a financial compensation scheme in place, available to transmission / distribution companies as well as to qualified customers. The issues that need to be resolved in this regard are the following:

1. Do all generators / synchronous compensators, irrespective of any other requirement, qualify as ancillary service providers?

2. Does the transmission / distribution companies, other customers, and other reactive power providers qualify as ancillary service providers?

3. How is the reactive capability of a service provider classified between the *mandatory service* component and the *ancillary service* component?

4. What is the likely structure of the reactive market and the payment mechanisms?

5. What are the issues involved in optimal procurement of reactive power ancillary services?

As seen above, the formation of a reactive power ancillary services market involves addressing several critical questions. These primarily arise due to the complex nature of reactive power itself as compared to real power markets. Some of the characteristics of a reactive power ancillary service market are as follows:

- Reactive power services need to be provided locally due to the technical problems associated with its transportation over lines

- The value of a reactive power service is location dependent, i.e. the worth of 1 MVAr of reactive power may be different at different buses

- Therefore strategic location of a reactive power service provider may lead to undue market advantages for some parties

- Due to the limited number of potential players in this market, there is an opportunity for market power

- The market will most likely be a monopsony[4].

[4] Monopsony is a market situation in which the product or service of several sellers is sought by only one buyer.

3.3 Synchronous Generators as Ancillary Service Providers

While reactive power ancillary services may be provided by many different parties such as generators, transmission / distribution companies, or even large customers, the importance of synchronous generators in providing the service remains a critical issue. This is due to their presence and their importance in the real power market, and which will also determine their role in the reactive power services.

In this sub-section we examine the reactive power generation capability of a synchronous generator. The power output of a synchronous generator is usually limited to a value within the MVA rating by the capability of its prime mover. When real power and terminal voltage are fixed its armature or field winding heating limits determine the reactive power generation from the generator. The armature heating limit is a circle (*Figure 12*) with radius $R_l=(V_tI_a)^{1/2}$ centered on the origin (15) and the field heating limit is a circle, centered at C_2, $(0, -V_t^2/X_S)$ with radius $R_2=V_tE_{af}/X_S$ (16) [21]. V_t is the voltage at the generator terminal bus, I_a is the steady-state armature current, E_{af} is the excitation voltage and X_S is the synchronous reactance. P and Q are real and reactive power generation from the machine, respectively.

$$P^2 + Q^2 \leq (V_t I_a)^2 \tag{15}$$

$$P^2 + \left(Q + \frac{V_t^2}{X_s}\right)^2 \leq \left(\frac{V_t E_{af}}{X_s}\right)^2 \tag{16}$$

The machine rating is the point of intersection of the two circles ('R' in *Figure 12*). When $P < P_R$, the limit on Q is imposed by the generator's field heating limit. While, when $P > P_R$ the armature heating limit imposes restrictions on Q. P_R is the real power corresponding to the machine rating.

If the generator is operating on the limiting curve, any increase in Q will require a decrease in P so as to adhere to the heating limits. Consider the operating point 'A' on the curve defined by (P_A, Q_A). If more reactive power is required from the unit, say Q_B, the operating point requires shifting back along the curve to point B (P_B, Q_B), where $P_B < P_A$. This signifies that the unit has to reduce its real power generation to adhere to field heating limits when higher reactive generation is demanded.

If the operating point lies inside the limiting curve, at (P_A, Q_{base}) then the unit can increase its reactive generation from Q_{base} up to Q_A without requiring re-adjustment of P_A.

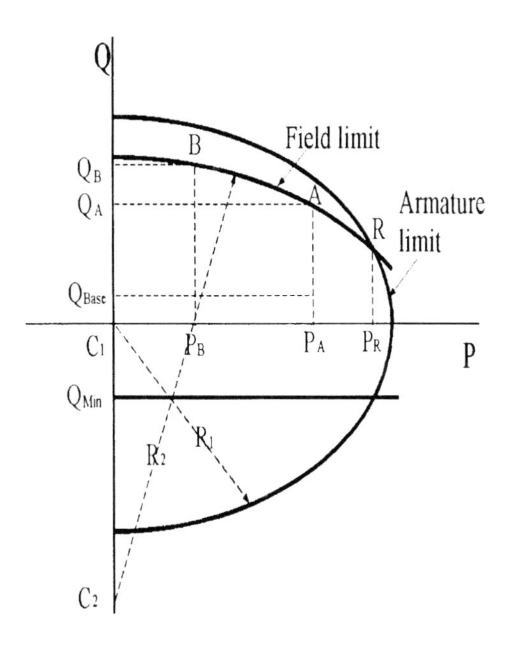

Figure 12. Synchronous generator capability curve

Figure 12 also includes a lower limit on Q, which restricts the unit operation in under-excited mode due to localized heating in the end region of the armature.

3.3.1 Cost of Reactive Power Production

Assume that a unit is operating at (P_A, Q_{base}). If the unit is required to increase its reactive power production from Q_{base} to Q_A, it will incur increased losses in the windings and hence an increase in its costs. This cost can be termed as the ***cost of loss component*** and is incurred by the synchronous generator with reactive power production (both lagging and leading). A typical reactive power versus loss plot has been shown in [22].

For reactive power production higher than Q_A, the generator has to reduce its real power generation in order to meet the constraints imposed by field or armature heating limits. It will hence incur revenue loss (RL), which needs to be compensated net of its cost savings from reduced generation, expressed as follows:

$$RL = \mu(P_A - P_B) - [C(P_A) - C(P_B)] \quad (17)$$

In (17), μ is the real power price and C(.) is the generation cost as a function of real power production. The term RL can also be called the ***opportunity cost component*** of reactive power generation for the generator.

The above two components are explained in *Figure 13*. However, it is to be noted that the nature of the plot shown is only figurative to illustrate the two components *i.e.* the cost of loss plot need not necessarily be a parabolic curve nor the other component be necessarily linear.

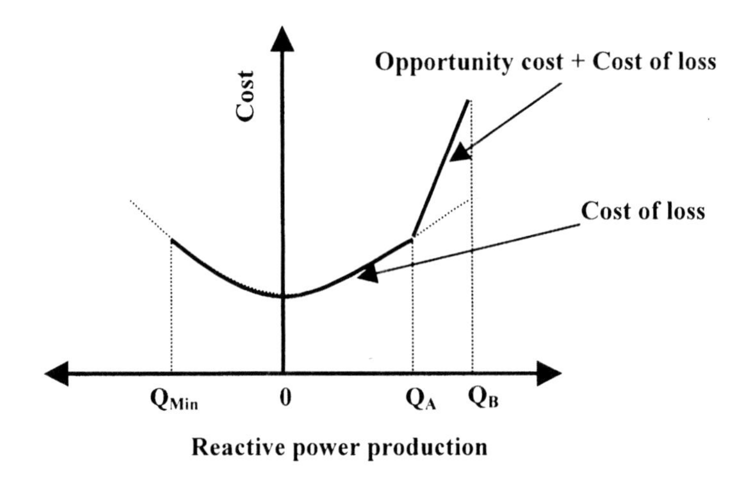

Figure 13. Reactive power production versus increased costs incurred by a synchronous generator

Understandably, the ISO will not be in a position to determine these two components for a generator since information on real power prices may not be known to the ISO if these are fixed through a bilateral contract and similarly the generator's cost of loss function may not be known. An appropriate option in such a case is to call for reactive bids from all generators [23, 24].

4. CONCLUDING REMARKS

This chapter attempts to examine the important area of ancillary services and how these are defined and handled in various electricity markets. We have discussed these services in the context of US regulatory changes and introduction of new guidelines by the North American Electric Reliability Council (NERC). These services have been compared with their counterparts in the UK, Australian and the Swedish electricity markets. We have seen that the balance services in Sweden and in other Nordic countries have worked quite well and UK has also initiated a move to introduce a similar service. On the other hand, the Nordic markets have no scheme for financial compensation for reactive power that exists in all other markets. Several complex issues are involved in the development of a proper reactive power market and those have also been highlighted.

REFERENCES

1. NERC While Paper, "Draft of Proposed Standards for Interconnected Operations Services", June 1999.
2. NERC Operating Policy-10 on Interconnected Operations Services, Draft-3.1, February 2000.
3. E. Hirst and B. Kirby, "Ancillary service details: Operating reserves", Oak Ridge National Laboratory Report, 1996, ORNL/CON-452.
4. B. Kirby and E. Hirst, "New black start standards needed for competitive markets", IEEE Power Engineering Review, Feb.'99, pp.9-11.
5. E. Hirst and B. Kirby, "Electric power ancillary services", Oak Ridge National Laboratory Report, 1996, ORNL/CON-426.
6. B. Kirby and E. Hirst, "Ancillary services costs for 12 US electric utilities", Oak Ridge National Laboratory Report, 1996, ORNL/Con-427.
7. The National Grid Company plc., "Ancillary services- an introduction", Second Edition, 1995.
8. National Electricity Market Management Company (Australia), "National electricity market ancillary services", Version-1.0, November 1999.
9. National Electricity Market Management Company (Australia), "Ancillary services costs- 98/99 financial year".
10. Svenska Kraftnät, "The Swedish Electricity Market and the Role of Svenska Kraftnät", Second Edition, 2000.
11. K. Walve, "Frequency control in the Nordic power system: Experiences and requirements", Proceedings of IEEE Power Engineering Society Winter Meeting 1999, pp. 559 - 560.
12. The National Grid Company plc., "An Introduction to Reactive Power: Ancillary Services- Reactive Contracts", June 1998.
13. N. H. Dandachi, M. J. Rawlins, O. Alsac, M. Prais and B. Stott, "OPF for Reactive Pricing Studies on the NGC System", IEEE Transactions on Power Systems, Vol.11, Feb.'96, pp. 226-232.
14. The National Grid Company plc., "NGC Reactive Market Report: Fourth Tender Round for Obligatory and Enhanced Reactive Power Services", November 1999.
15. The National Grid Company plc., "NGC Reactive Market Report: Fifth Tender Round for Obligatory and Enhanced Reactive Power Services", May 2000.
16. New York Independent System Operator Ancillary Services Manual, 1999.
17. National Electricity Market Management Company (Australia), "National Electricity Market Ancillary Services", November 1999.
18. National Electricity Market Management Company (Australia), "Generator Code Reactive Obligations", November 1988.
19. National Electricity Market Management Company (Australia), "Operating Procedure: Ancillary Services", Document Number SO_OP3708.
20. FINGRID OY Main Grid Service Conditions 1, January 1999.
21. E. Fitzgerald, C. Kingsley, Jr., S. D. Umans, Electric Machinery, McGraw-Hill, 1992.

22. J. B. Gil, T. G. S. Roman, J. J. A. Rios and P. S. Martin, "Reactive power pricing: A conceptual framework for remuneration and charging procedures", IEEE Transactions on Power Systems, May 2000.
23. K. Bhattacharya and J. Zhong, "Reactive Power as an Ancillary Service", IEEE Transactions on Power Systems, forthcoming.
24. J. Zhong, K. Bhattacharya and Jaap Daalder, "Reactive power as an ancillary service: Issues in optimal procurement", Proc. of IEEE-PES / CSEE International Conference on Power System Technology, POWERCON 2000, Vol.2, pp.885-890.

Chapter 6

Reliability and Deregulation

1. TERMINOLOGY

Power system reliability is the field within power engineering that treats the ability of the power system to perform its intended function. This is a rather wide field as "its intended function" can be interpreted in many different ways; from "generating and transporting electrical energy" to "giving customers the best value for their money". In practice, power system reliability concerns the development of methods to stochastically predict the number and duration of interruptions of the electricity supply. The term "prediction" is used here in a completely different way as in daily life. No reliability analysis technique will ever be able to come to conclusions such as "an interruption will take place in a certain small village in the southern part of The Netherlands on 21 July at 10:20 in the morning, lasting 39 minutes". Instead, the kind of result from a reliability analysis is more like "a rural customer in The Netherlands can expect 0.2 interruptions during the month of July, 50% of which are expected to take place between 8 am and 5 pm".

Within power system reliability, three terms play an important role: "failure", "outage" and "interruption". Whereas the meaning of these three terms is similar in daily life, the difference between them is sometimes essential for an understanding of power system reliability. The definition of these terms requires some additional attention.

- *Failure*: The term failure is used in the general meaning of the term: a device or system which does not operate as intended. Thus we can talk about a failure of the protection system to clear a fault, but also of the failure of a transformer, and even about the failure of the public supply.

- *Outage*: An outage is the removal of a primary component from the system, *e.g.* a transformer outage or the outage of a generator station. A failure does not necessarily have to lead to an outage, *e.g.* the failure of the forced cooling of a transformer. And the other way around, an outage is not always due to a failure. A distinction is therefore made between "forced outages" and "scheduled outages". The former are directly due to failures, the latter are due to operator intervention. Scheduled outages are typically to allow for preventive maintenance, but also the aforementioned failure of the forced cooling of a transformer could initiate the scheduling of a transformer outage.

- *Interruption*: An interruption is the situation in which a customer is no longer supplied with electricity due to one or more outages in the supply. In reliability evaluation the term interruption is used as the consequence of an outage (or a number of overlapping outages), which is in most cases the same as the definition used in the power quality field (a zero-voltage situation).

- The term *blackout* will be used for a long interruption covering a very large number of customers. There is no strict boundary, but with a blackout one thinks of several million of customers, a large metropolis, or a large part of a country, being affected for several hours. From a theoretical viewpoint there is no difference between an interruption and a blackout. But publicity and political consequences of a blackout are much more than those of a normal interruption.

2. RELIABILITY ANALYSIS

A reliability analysis study, as mentioned in the previous section, gives a stochastic prediction of number and duration of interruptions. Each such study consists, explicitly or implicitly, of the following steps:

1. ***Describe the supply system to be analyzed:*** Both the primary system and the protection system need to be considered, even if the latter will not be part of the actual study. At this stage of the study, the realm of the analysis will be determined. Since it is not possible to consider the whole power system in all its detail, the relevant parts of it need to be selected. This selection process is very much application dependent as well as

determined by personal preferences. Next to a description and limitation of the system to be studied, this part of the study also will result in the "*interruption criterion*". Examples of this will be given below.

2. ***Define the stochastic components:*** The physical system is translated in a mathematical model by splitting it into so-called stochastic components that behave in a well-defined stochastic manner. The choice of stochastic components suffers from the same problems as many deterministic modeling studies: the model needs to be sufficiently detailed to obtain accurate results; but the model needs to be simple enough to enable these results to be obtained within a reasonable time. Although the choice of stochastic components does not limit the choice of component model or calculation method, the number of components is equally important in determining calculation speed and accuracy.

3. ***Choose a component model:*** Most reliability studies use the simplest model available, a two-state memory-less model, but more detailed models certainly need to be considered. The availability of faster computers makes the application of more accurate models possible. The choice of such a more accurate component model will limit the choice of calculation methods in the next stage of the analysis. The choice in component model may also be limited by the availability of component data. If, for example, no data is available on component ageing, it may be not feasible to include ageing in the component model.

4. ***Choose a calculation method:*** When the basic simple component model is used, as is normally the case, several competing calculation methods are available. The balance between accuracy and speed/complexity of the method again rules this choice. With the choice of more accurate component models, the choice of calculation method may become trivial, with the Monte-Carlo simulation being the only feasible method remaining for very complicated component models.

5. ***Perform the calculations:*** Having gone through the previous steps, performing the actual calculations may appear rather trivial, but it rarely is. Doing a reliability analysis is not just a matter of obtaining the results. The interpretation of the results is very important and rarely trivial. The results should *e.g.* not contradict the assumptions made during the modeling process.

2.1 The Interruption Criterion

As mentioned before, the aim of a reliability analysis is to assess the performance of the power system. For such an assessment it is essential to define "performance": against which criterion the performance of the system is compared. This part of the study can easily lead to a deeply philosophical discussion. But, no matter how interesting is such a discussion, it can better be avoided at this stage. What is needed for the reliability analysis study is a well-defined mathematical criterion to distinguish between an interruption-state and a non-interruption state of the system. This criterion is referred to as the "interruption criterion". For each system state or for each event, the interruption criterion can be used to determine if this state or event should be counted as an interruption or not. In most studies the interruption criterion is simple, but for more detailed studies, especially for Monte Carlo simulation, the definition of the interruption criterion becomes an important part of the modeling effort. It is recommended to spend some time on defining the interruption criterion for each study. Some simple examples of interruption criteria are given below. Note that these are just examples, and certainly not the only possibilities.

- In a generation reliability study, a state is an interruption state if the generator capacity is less than the load demand. Note that there is only one interruption criterion for the whole system. Each customer is equal at this level.

- In a transmission reliability study a state is an interruption state for a given transmission substation if the maximum power that can be transported to this substation is less than the load demand. For transmission reliability studies, each substation has its own interruption criterion, thus its own reliability.

- In a transmission security study an event is an interruption event if both the initial system-state and the resulting system-state are not interruption states, and the transient phenomenon due to the event leads to tripping of generators and/or load.

- In a distribution reliability study a state is an interruption state for a given customer if the voltage at the customer terminals is zero.

- In a distribution power quality study an event is an interruption event for a given device if it leads to the (voltage magnitude) event at the device terminals to exceed a certain magnitude and duration.

2.2 Stochastic Components

For a reliability evaluation study, the power supply system is split into stochastic components. The choice of components is rather arbitrary: the whole transmission system might be one component, but a single relay could be several components. Each component can be in at least two states: healthy and non-healthy, the latter often referred to as the outage state. For a two-state component, two events can occur: the transition from the healthy to the non-healthy state, an outage or failure event; and the reverse transition (i.e. from the non-healthy to the healthy state), the repair or restore event.

The system-state is a combination of all event states; if the state of one of the component changes, the system state changes. The system-state for a system with N components can be thought of, as a vector of rank N. The value of each element is the state of the corresponding component. An event is a transition between two system states, due to the change in state of one or more components.

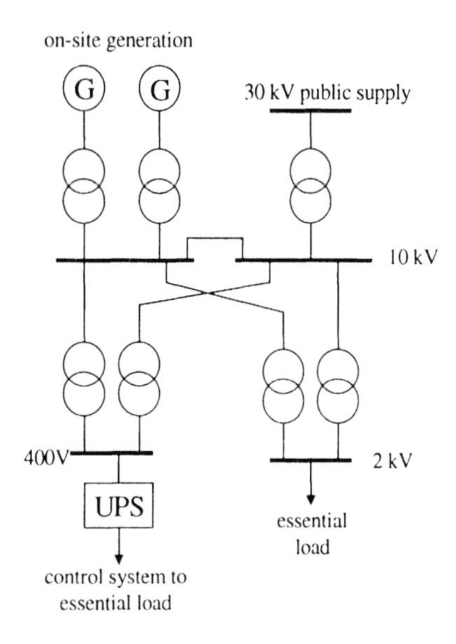

Figure 1. Example of a power system

Consider, as an example, the power system shown as a one-line diagram in *Figure 1*. A possible choice of stochastic components is as follows:

- each of the on-site generators plus generator transformer
- the public supply
- the 30/10 kV transformer

- each of the 10 kV busses
- the connection between the two 10 kV busses
- each of the 10 kV/400 V connections
- each of the 10/2 kV connections
- the uninterruptible power supply (UPS)

In a more detailed study one may decide to split up each transformer connection into:
- transformer internal fault
- transformer terminal fault
- transformer relay
- switchgear on primary side of transformer
- switchgear on secondary side of transformer
- cable on primary side of transformer
- cable duct on secondary side of transformer

The other components may also be subdivided in a similar amount of detail. When choosing the stochastic components it is important to realize that electrical engineers are trained to think in terms of circuit elements. The choice of components above clearly reflects this. However, secondary components often significantly effect number and duration of interruptions, and they therefore need to play an important role in a reliability analysis study. Events like stuck-breaker and protection mal-trip need to be seriously considered in any study. Unfortunately, some of the calculation methods make it difficult to include these. Also, failure data for these events is not always available.

2.3 Component Models

2.3.1 The Basic Component Model

Two quantities are normally used to describe the behavior of a stochastic component: the failure rate and the (expected or average) repair time. The meaning of the term "expected repair time" is obvious: the expected value of the time the component resides in the non-healthy state. The failure rate λ gives the average probability that the component will fail in the next small period of time. In equation form:

$$\lambda = \lim_{\Delta t \to 0} \frac{\Pr(\text{failure in period } \Delta t)}{\Delta t} \qquad (1)$$

For components representing primary parts of the power system, which accounts for the majority of the components in most studies, the term outage rate might be used. Here we shall use the general term failure rate.

The expected time to failure T is the reciprocal of the failure rate:

$$T = \frac{1}{\lambda} \tag{2}$$

The repair rate μ is the reciprocal of the expected time to repair R:

$$\mu = \frac{1}{R} \tag{3}$$

Note that expected time to failure can be defined in a similar way as the expected repair time, and the repair rate similarly as the failure rate.

The availability of the component is the probability to find the component in the healthy state:

$$P = \frac{T}{R+T} \tag{4}$$

The unavailability is the probability that the component is in the non-healthy state:

$$Q = \frac{R}{R+T} \tag{5}$$

Example: A distribution company operates 7500 distribution transformers. Over a period of 10 years, 140 of these transformers fail for various reasons. A small fraction of them can be repaired, but most failures require replacement with a spare transformer. Records have been kept of the repair or replacement time needed. Adding all these for the 140 failures gives a total of 7360 hours. From these observation data, the values of the above parameters are obtained:

$$\lambda = \frac{140}{10 \times 7500} = 0.0019\ \text{yr}^{-1}$$

$$T = \frac{1}{0.0019} = 530\ \text{yr}$$

$$R = \frac{7360}{140} = 52.6\ \text{h} = 0.006\ \text{yr}$$

$$\mu = \frac{1}{R} = 167\ \text{yr}^{-1}$$

$$P = \frac{530}{0.006 + 530} = 0.999989$$

$$Q = \frac{0.006}{0.006 + 530} = 6\ \text{min/yr}$$

This can be interpreted in normal words, as follows:

- Each transformer has a probability of 0.0019 to fail in the coming year. In the whole population, 14 transformers are expected to fail each year.
- After such a failure, the repair or replacement of the transformer is expected to last 52.6 hours.
- Each transformer will be out of operation for, on average, 6 minutes per year.

Note that we have used past-performance data to predict future behavior. This is the basis for all reliability analysis: the assumption that the average performance in the past gives the expected behavior for the future.

2.3.2 Multi-State Components

In the previous section, it was assumed that a stochastic component resides in one of two states: one healthy state and one non-healthy state. Where this model is sufficient for many components, in some cases multi-state models are needed. An example of a multi-state component model is shown in *Figure 2*. Next to the "healthy state", there are three "non-healthy states": faulted, repair and maintenance. (One can argue whether "maintenance" is a healthy or a non-healthy state, but such a discussion is not relevant here.) From the healthy state, the component can reach the maintenance-state and the faulted state. If the component represents part of the primary system, *e.g.* a power transformer, maintenance will lead to a planned outage. The transition from the healthy to the maintenance-state is a planned transition. It is the system operator who decides to take the transformer out of operation for maintenance. The transition from the healthy state to the faulted state is a forced transition, it is outside of the control of the system operator. The faulted state is normally a short circuit after which the protection brings the component in the repair-state. Some faults, *e.g.* failure of the cooling for a transformer, may bring the component in the maintenance-state.

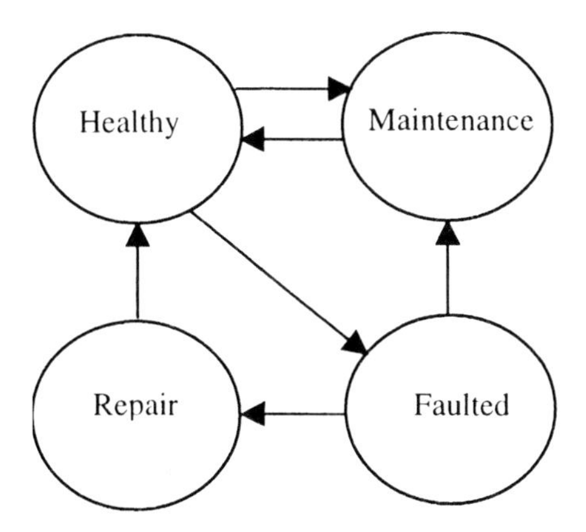

Figure 2. Four-state component model

It may be appropriate in some studies to split the faulted state in two: faults that require repair and faults that require maintenance, as shown in *Figure 3*. The processing of the two faulted state may be completely different: for example, the "faulted 1 state" is represented as a short circuit in the primary system, where the "faulted 2 state" enables normal operation of the component until it is taken out of operation for maintenance. The power rating of the component (*e.g.* the power transformer) may be less in the "faulted 2 state" than in the "healthy state". Such so-called "derated states" are commonly used in the modeling of power generators.

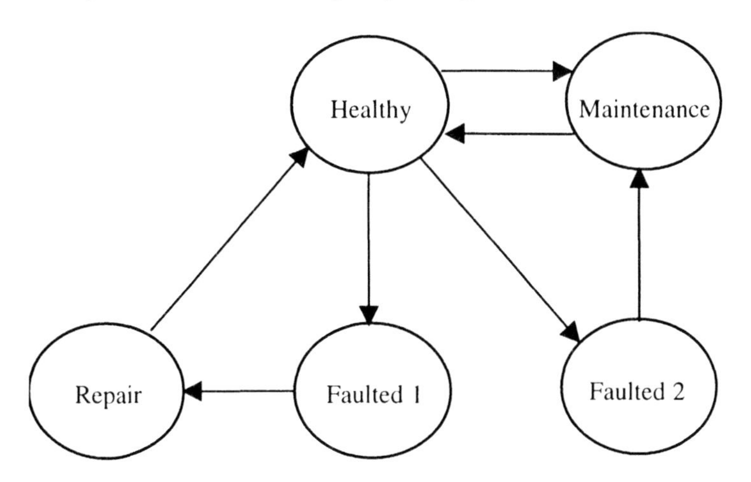

Figure 3. Five-state component model obtained by splitting the "faulted state" in *Figure 2*.

2.3.3 The General Component Model

Describing a stochastic component by means of two quantities (*e.g.* failure rate and repair time) is a gross simplification of the actual situation. Still this model is used in most studies for several reasons beyond the scope of this text. In a more general component model a probability distribution is defined for the time until a given transition (or alternatively for the time the component stays in each state). There is thus one probability distribution function for the repair time (the time in the non-healthy state) and one for the lifetime (the time in the healthy state). Let T be the life (expected time to failure) of the component. The probability distribution function of life time $F(t)$ is the probability that the component fails before it reaches an age t:

$$F(t) = \Pr\{T \le t\} \tag{6}$$

The probability density function is the derivative of the probability distribution function:

$$f(t) = \frac{dF}{dt} = \lim_{t \to 0} \frac{\Pr\{t < T \le t + \Delta t\}}{\Delta t} \tag{7}$$

The probability density function is a measure for the probability that the component will fail around an age t: The failure rate $\lambda(t)$ is defined as the probability that the component fails soon after the age t assuming that it has not failed before:

$$\lambda(t) = \lim_{\Delta t \to 0} \frac{\Pr\{T \le t + \Delta t \mid T > t\}}{\Delta t} \tag{8}$$

The failure rate can be calculated from the probability density function and the probability distribution function by using the following expression:

$$\lambda(t) = \frac{f(t)}{1 - F(t)} \tag{9}$$

Using the negative-exponential distribution for both time-to-failure and repair time, results in the before-mentioned basic component model (see Section 2.3.1). This basic component model is the most commonly used model, but more accurate models using more accurate distribution functions are also possible [2, 4, 5, 8], and their use should be encouraged.

2.3.4 Electric Circuit Representation of the System

In the previous sections the stochastic modeling of the system was discussed. The stochastic behavior of the power system is represented by a finite number of stochastic components, each with a finite number of states and transitions. The difference between various models is in the number of states and the distribution functions used for the transitions. But the stochastic model alone cannot fully describe the behavior of the system, as there remains a deterministic aspect, i.e. the circuit theory model of the system. The primary part of the power system is represented as a number of impedances, as in any other power-system analysis technique (load flow, short-circuit currents, etc.). The main difference with reliability analysis is that the electric circuit model of the system changes stochastically. Each transition in the stochastic component model may be associated with a change in the electrical circuit. Some examples are already mentioned in Section 2.3.2 above: during maintenance the component is out of operation; in circuit-theory terms: the impedance is replaced by an open circuit. During a fault the component is represented by a short-circuit, *etc.*

The amount of detail needed again depends on the type of study. For a generation reliability study, the power system model is limited to one node. For a transmission reliability study, component overload may play an important role, so that each connection should be represented by an impedance. When security and voltage dips are to be considered a more detailed circuit-theory model may be needed, in which *e.g.* the machines are modeled as a variable impedance [3].

Note that different stochastic components may affect the same impedance, and the other way around. An example of this is shown in *Figure 4*. A transformer connection consisting of a circuit breaker, an underground cable, the transformer, and a fuse on low-voltage side of the transformer, is represented by five stochastic components: the circuit breaker; the cable; the transformer; the transformer protection; and the current-limiting fuse. The stochastic events shown in the figure, all have the same effect on the electric circuit: removal of the impedance. As the effect on the system is the same, all these events are often modeled as one stochastic event. This greatly simplifies the complexity of the calculations, but it may lead to a certain loss in accuracy. We will see below that, for the basic component model using the exponential distribution, such an aggregation can be done without loss of accuracy.

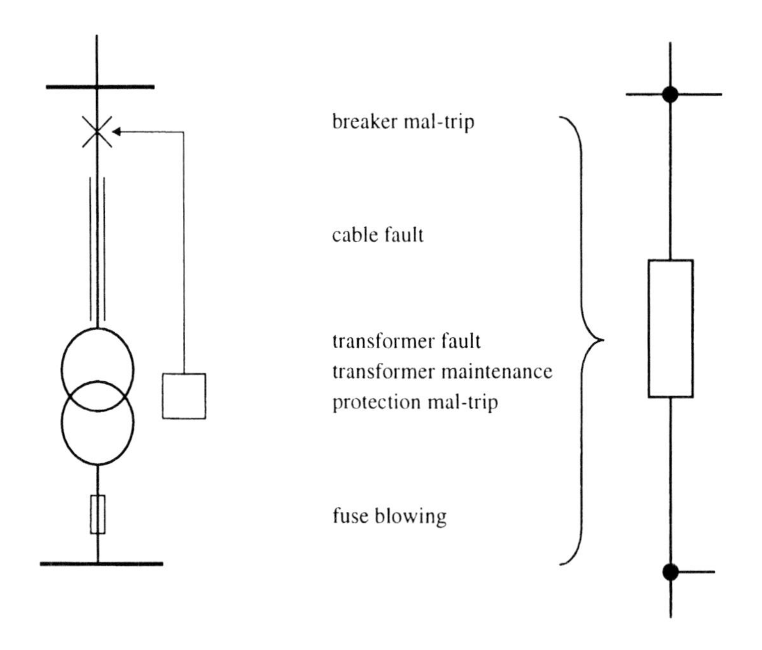

Figure 4. Different stochastic components affecting the same impedance

2.4 Calculation Methods

When the component models and their relation to the electrical circuit have been decided, it is time to choose a calculation method. There are a myriad of calculation methods available in books and papers, but they can be grouped (with most likely some exceptions not treated here) into the following basic methods:

- The "*network model*", to be discussed in Section 3, is the most simple method available. It can only be used for the basic component model using exponential distributions. It also places severe limitations on the system model. Despite these limitations it remains a commonly used method. Even most other methods use the exponential distribution for component models. Other methods do enable more accurate system models, but actually implementing them is certainly not trivial. Limitations in computer power, development time, component data, and engineering insight, make that other methods often result in similar conclusions as obtained from the network model.

- A *Markov model* is state-based method that enables much more details in the system model. In principle, this method does not put any limitation on the system model, only on the component model. Each transition in the component model needs to be exponentially distributed, and independent of the previous state of the component. Under these assumptions it is possible to describe the whole system through a large number of states, with only exponentially distributed transitions. The state probability, state duration, and the transition frequency can be analytically calculated. The exact results can be obtained through the inversion of a matrix with order equal to the number of system states. As this number is extremely large such an inversion is in practice not possible, and approximation methods are needed. One possibility is to obtain recursive expressions in which the probability of a state is obtained from the probability of a number of more-likely states [7]. The main limitation of Markov models is the restriction to exponential distributions, which gives incorrect results for the distribution of the duration of interruptions. Markov models can therefore not be used to determine the expected costs of interruptions. So-called semi-Markov models enable the use of non-exponential distributions in Markov-like component and system models. The so-called Weibull-Markov model enables the use of a generalised exponential distribution: the so-called "Weibull distribution", [4, 5].

- *Monte-Carlo simulation methods*, do not perform any analytical calculations, but instead simulate the stochastic behavior of the system. Any component model, with any distribution, and any system detail, can be included in a Monte-Carlo simulation [1]. The only limitations on the model complexity are computer power, component data, and engineering insight. Also, simulation techniques tend to give less insight in the results than analytical methods.

- *Renewal theory* is a set of mathematical techniques that enable analytical calculations for component models with any distribution. The resulting expressions are unfortunately of such a complexity that only simple systems can be studied. However, under certain assumptions it is possible to use results from renewal theory in power system reliability analysis [15].

In this chapter, only the network model will be discussed further. For details on other calculation methods, the reader is referred to the literature. Several basic texts and overview papers can be found in the list of further reading at the end of this chapter.

3. THE NETWORK MODEL

3.1 Stochastic Networks

When using the so-called network approach, the system is modeled as a "stochastic network". The stochastic behavior of the system is represented graphically by means of a number of network blocks, connected in parallel or in series. Each block refers to a stochastic component in the system. The model is such that the system is healthy (i.e. the supply is available) as long as there is a path through the network. This graphical character of the method makes it very suitable to get an overview of the reliability of the system. An additional advantage of the network approach is the similarity with the electrical network. Electrically parallel components are often modeled as a parallel connection in the stochastic network. An electrical series connection in most cases results in a stochastic series connection.

When the reliability is quantified by using a stochastic network, a number of mathematical approximations are needed. The calculations assume that the repair time and the lifetime are exponentially distributed for all components.

Each block (network element) is characterized through an outage rate λ and an expected repair time r. We further define the "availability" P and the "unavailability" Q for each element:

$$P = 1 - \lambda r \tag{10}$$

$$Q = \lambda r \tag{11}$$

Consider as an example the transmission system shown in *Figure 5*. In a first stage of the study it is decided to only include failure of the generators and failure of the transmission lines. Substation-related failures are not included in the study. A possible stochastic network for the supply to load 1 is shown in *Figure 6*. The numbers refer to the numbers in *Figure 5*. The assumptions made in obtaining the network diagram are, among others, that one generator and one transmission line are sufficient to supply the load. Overload of transmission lines, and load shedding due to lack of generation is not included here.

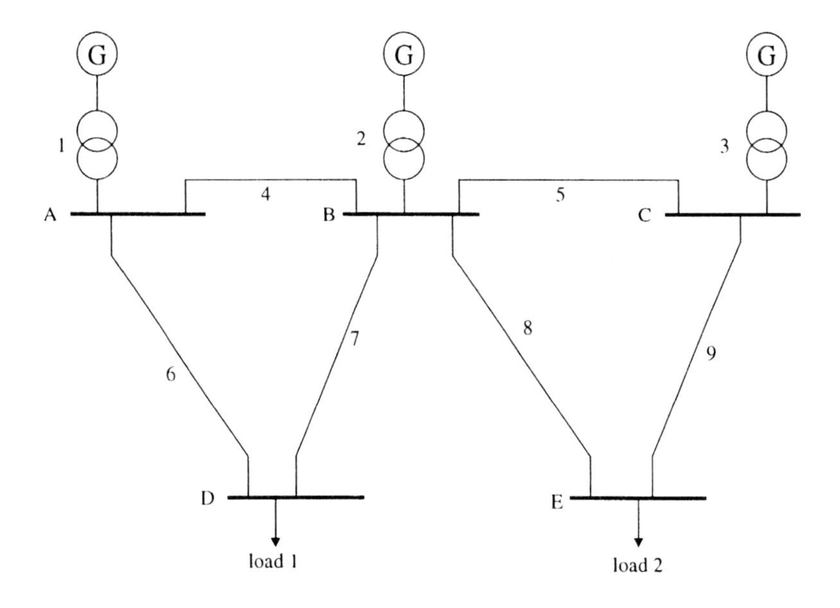

Figure 5. Example of transmission system

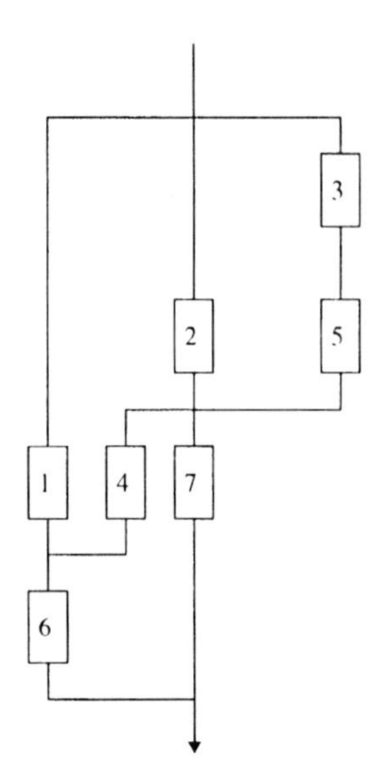

Figure 6. Network model for the system shown in *Figure 5*

3.2 Series and Parallel Connections

Consider the series connection of two stochastic components with outage rate λ_1 and λ_2 and repair time r_1 and r_2 as shown in *Figure 7*. We want to derive expressions for outage rate λ_s and repair time r_s of the series connection; so that the series connection can be replaced by one equivalent component.

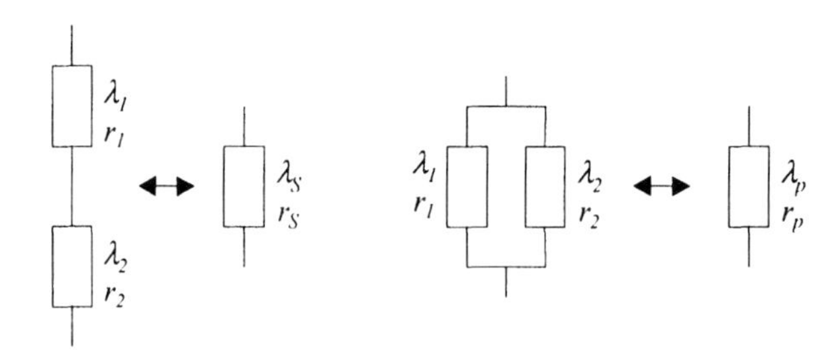

Figure 7. Stochastic series connection (left) and stochastic parallel connection (right)

A series connection fails when either of the components fails. The outage rate for the series connection is thus the sum of the outage rates of the components:

$$\lambda_S = \lambda_1 + \lambda_2 \tag{12}$$

The series connection is not available when one of the components is not available, giving for the unavailability of the series connection:

$$Q_S = Q_1 + Q_2 \tag{13}$$

Using the definition of unavailability, (11) , gives an expression for the equivalent repair time of the series connection:

$$r_S = \frac{\lambda_1 r_1 + \lambda_2 r_2}{\lambda_1 + \lambda_2} \tag{14}$$

For n components in series, the following expressions can be derived:

$$\lambda_S = \sum_{i=1}^{n} \lambda_i \tag{15}$$

$$r_S = \frac{\sum_{i=1}^{n} \lambda_i r_i}{\sum_{j=1}^{n} \lambda_j} \tag{16}$$

In deriving these expressions a number of assumptions have been made, all coming back to the system being available most of the time, thus $\lambda r << 1$.

A parallel connection, as shown on the right-hand side of *Figure 7*, fails when one of the components is not available and the other one fails: thus when 1 is unavailable and 2 fails or when 2 is unavailable and 1 fails. The outage rate of the parallel connection is thus:

$$\lambda_p = Q_1 \lambda_2 + Q_2 \lambda_1 = \lambda_1 \lambda_2 (r_1 + r_2) \tag{17}$$

The parallel connection is not available when both components are not available. The unavailability of the parallel connection is:

$$Q_p = Q_1 Q_2 \tag{18}$$

The repair time of the parallel connection is obtained from (17) and (18):

$$r_p = \frac{r_1 r_2}{r_1 + r_2} \tag{19}$$

The same process can be repeated several times, resulting in the following general expressions for a system consisting of n components in parallel:

$$\lambda_{pn} = \prod_{i=1}^{n} \lambda_i r_i \sum_{j=1}^{n} \frac{1}{r_j} \tag{20}$$

$$\frac{1}{r_{pn}} = \sum_{j=1}^{n} \frac{1}{r_j} \tag{21}$$

3.3 Minimum Cut-Sets

A commonly used method for performing calculations in stochastic networks is the so-called "method of minimum cut-sets". A "cut-set" is a combination of components whose outage would lead to an interruption. In the stochastic network of *Figure 6* the combinations {1,2,4,5} and {6,7} are examples of cut-sets. A cut-set is a "minimum cut-set" if the removal of any one of the components from the cut-set would make it no longer a cut-set. In other words: if one of the components of a minimum cut-set is repaired, the supply is restored. In *Figure 6* the cut-set {1,2,4,5} is no minimum cut-set because repair of component 4 does not restore the supply. The cut-set {6,7} is a minimum cut-set because both repair of component 6 and repair of component 7 will restore the supply. For each stochastic network there are a limited number of minimum cut-sets. Finding all the minimum cut-sets is a first step in the method of minimum cut-sets.

For the stochastic network in *Figure 6* the following minimum cut-sets can be found:

- {6,7}
- {1,2,3}
- {1,2,5}
- {1,4,7}
- {2,3,4,5}
- {2,4,5,6}

The system fails when any of these 6 minimum cut-sets fails. The system can thus be viewed as a series connection of 6 minimum cut-sets. A minimum cut-set fails when all its components fail, and is thus a parallel connection of its components. In other words: knowing the minimum cut-sets, the system can be transformed into a series connection of parallel connections. The result for the current example is shown in *Figure 8*. Using the expressions for parallel and series connections derived before, the failure rate and repair time of the whole system can be obtained easily.

Note that this calculation method is not restricted to stochastic networks. When the minimum cut-sets can be obtained in any other way, *e.g.* directly from the description of the system, the same calculations can be applied.

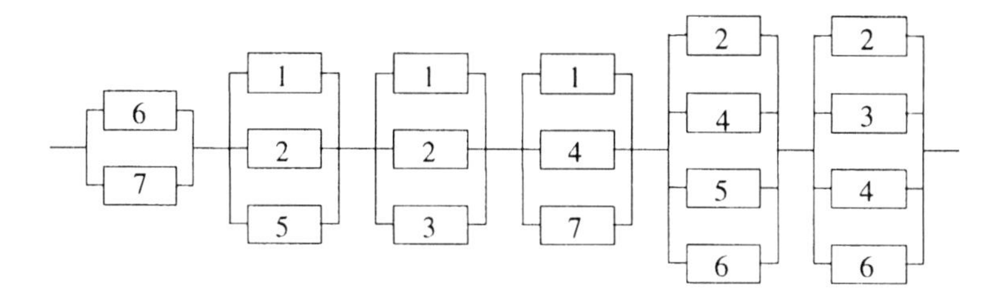

Figure 8. Series connection of parallel connection for the system shown in *Figure 6*,

4. RELIABILITY COSTS

To consider interruptions of the supply in the design and operation of power systems, the inconvenience due to interruptions needs to be quantified one way or the other. The term inconvenience is a rather vague and broad term. Any serious quantification requires a translation of all inconvenience into amount of money.

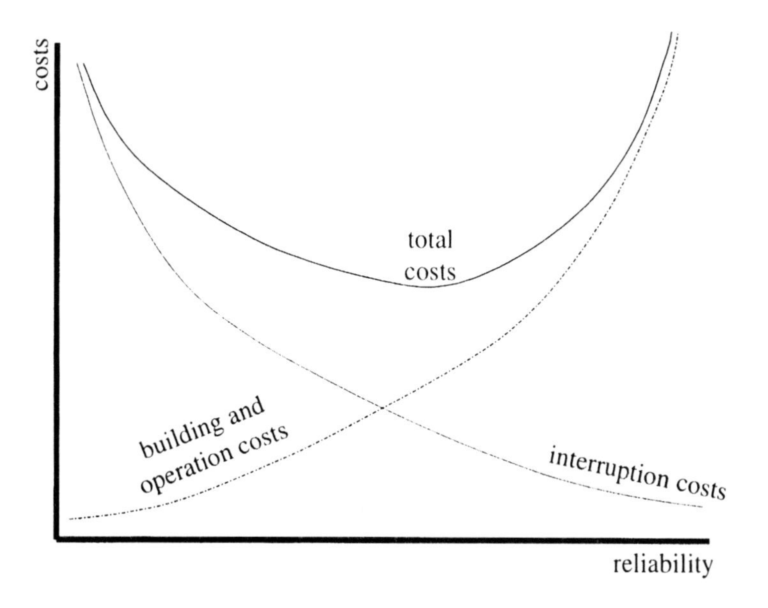

Figure 9. Costs versus reliability: costs of building and operation; costs of supply interruptions and total costs

Many publications on costs of interruption show a graph with costs against reliability. Such a curve is reproduced in *Figure 9*. The idea behind this curve is that a more reliable system is more expensive to build and to operate, but the costs of interruption (either over the lifetime of the system, or per year) are less. The total costs will show a minimum, which corresponds to the optimal reliability. Even if we assume that both cost functions can be determined exactly, the curve still has some serious limitations. *Figure 9* should only be used as a qualitative demonstration of the trade-off between costs and reliability.

- Additional investment does not always give a more reliable system. An increase in the number of components could even decrease the reliability.
- Reliability is not a single-dimensional quantity. Both interruption frequency and duration of interruption influence the interruption costs.
- There is no sliding scale of reliability and costs. The system designer can choose between a limited number of design options, sometimes there are just two options available. The choice becomes simply a comparison of advantages and disadvantages of the two options.
- The two cost terms cannot simply be added. One term (building and operational costs) has a small uncertainty, the other term (interruption costs) has a large uncertainty due to the uncertainty in the actual number and duration of interruptions. A more detailed risk analysis is needed than just adding the expected costs.
- There is a phenomenon called "progress" which makes that engineering systems become better, not necessarily cheaper. If *Figure 9* had been applied in the 18th century, power systems would never have been built.

The cost of an interruption consists of a number of terms. Each term has it's own difficulty in being assessed. Again simply adding the terms to obtain the total costs of an interruption is not the right way, but due to lack of alternatives it is often the only feasible option.

***Direct costs**:* These are the costs that are directly attributable to the interruption. The standard example for domestic customers is the loss of food in the refrigerator. For industrial customers the direct costs consist, among others, of lost raw material, lost production, and salary costs during the non-productive period. For commercial customers the direct costs are the loss of profit and the salary costs during the non-productive period. When assessing the direct costs, one has to be watchful of double counting. One should at first subtract the savings made during the interruption. The obvious saving is in the electricity costs, but for industrial processes there is also a saving in use of raw material.

Indirect costs: The indirect costs are much harder to evaluate, and in many cases not simply to express in amount of money. A company can lose future orders when an interruption leads to delay in delivering a product. A domestic customer can decide to take an insurance against lost of freezer contents. A commercial customer might install a battery backup. A large industrial customer could even decide to move a plant to a country with less supply interruptions. The main problem with this cost term is that it cannot be attributed to a single interruption, but to the (real or perceived) reliability as a whole.

Non-material inconvenience: Some inconvenience cannot be expressed in money. Not being able to listen to the radio for two hours can be a serious inconvenience, but the actual costs are zero. In industrial and commercial environments, the non-material inconvenience can also be big without contributing to the direct or indirect costs. A way of quantifying these costs is to look at the amount of money a customer is willing to pay for not having this interruption.

To evaluate the costs of supply interruptions, different methods have been proposed. For large industrial and commercial customers an inventory of all the direct and indirect costs can be made, and this can then be used in the system design and operation. Even for small customers such a study could be made, *e.g.* to decide about the purchase of equipment to mitigate interruptions. However for small and domestic customers it is often the non-material inconvenience that has a larger influence on the decision than the direct or indirect costs. For a group of customers, such an individual assessment is no longer possible. The only generally accepted method is the large survey among customers. Customers get asked a number of questions. Based on the answers the average costs of interruption are estimated. Alternatively, the customers are asked about their willingness to accept higher electricity prices in return for a reduced interruption frequency or duration. Some references to studies performed are included in the list of further reading at the end of this chapter.

To quantify the costs of an interruption, again different methods are in use, where it is unfortunately not always clear from the context which method is actually used.

Costs per interruption: For an individual customer the costs of an interruption of duration d can be expressed in dollars or in any other currency. There is no confusion possible about this. For simplicity, we neglect the fact that the costs not only depend on the duration but also on many other factors. The costs per interruption can be determined through an inventory of all direct and indirect costs.

Costs per interrupted kW: Let $C_i(d)$ be the costs of an interruption of duration d for customer i, and L_i the load of this customer when there would not have been an interruption. The costs per interrupted kW are defined as:

$$\frac{C_i(d)}{L_i} \tag{22}$$

and are expressed in \$/kW. For a group of customers experiencing the same interruption, the costs per interrupted kW are defined as the ratio of the total costs of the interruption and the total load in case there would not have been an interruption:

$$\frac{\sum_i C_i(d)}{\sum_i L_i} \tag{23}$$

Costs per kWh not delivered. In many studies the assumption is made that the cost of an interruption is proportional to the duration of the interruption. The cost per kWh not delivered is defined as:

$$\frac{C_i(d)}{dL_i} \tag{24}$$

and is constant under the assumption. The cost per kWh is expressed in \$/kWh. The term "value of lost load" is sometimes used for the cost per kWh not delivered averaged over all customers.

Costs of interruption rated to the peak load: A problem in surveys is that the actual load of individual customers in case there would not have been an interruption, is often not known. One should realize that surveys consider hypothetical interruptions, rarely actual ones. For industrial and commercial customers the peak load is much easier to obtain, as it is typically part of the supply contract. The cost of an interruption can be divided by the peak load, to get a value in \$/kW. Some care is needed when interpreting this value, as it is not the same as the cost per kW interrupted (also in \$/kW).

Costs per interruption rated to the annual consumption: For domestic customers it is easier to obtain the annual consumption than the peak load. Rating the cost of an interruption to the annual consumption gives a value in \$/kWh. Note that this has no relation to the costs per kWh not delivered.

5. HIERARCHICAL LEVELS

The power system is often divided into three functional parts, each with its own specific design and operation problems and solutions:
- generation
- transmission
- distribution

In the reliability analysis a similar distinction is made between three so-called hierarchical levels of reliability:
- Level-I: generation
- Level-II: transmission
- Level-III: distribution

The concept of hierarchical levels remains an approximation, as most classifications. Also the concept of hierarchical levels has been developed for the large public supply system in industrialized countries. For developing countries, for small insular systems, and for industrial power systems different thoughts might be needed. Despite the shortcomings of the classification in hierarchical levels, it still gives a good insight into the subject.

5.1 Generation Reliability

For most industrialized countries, outages of generators have no influence whatsoever on the interruption frequency nor on the supply availability experienced by a customer. Thus for a customer, level-I reliability studies do not appear very important. This conclusion is correct for an existing, well-planned and well-operated power system. But in the planning stage, level-I studies are extremely important. In modern power systems, generation of power takes place at the highest voltage level, thus a lack of generation capacity becomes immediately a national or even international problem. Such a situation should be avoided as much as possible. Because a suitable reserve in generation capacity has been planned and is available during operation, the customer does not have to worry about lack of generation anymore.

The rule that the total generation capacity in a power system should exceed the annual peak load is likely to be the most important planning criterion in power systems. Planning and building of large power stations takes between 5 and 10 years, thus decisions about this have to be made several years in advance. The most basic level I reliability study is to calculate the probability that the available generation capacity is less than the annual peak load in a certain year (*e.g.* 7 years ahead of the decision date). The input data for such a study consists of the expected annual peak load, the capacity of each generator unit and its forced unavailability. The *forced unavailability* is the fraction of time during which a unit is not available due to forced outages, i.e. during which it is in repair. The assumption to be made is that the probability that the unit is not available during the annual peak is equal to the forced unavailability. This gives sufficient information to calculate the probability that the available capacity is less than the annual peak load. This probability is called the "*loss of load expectation*" (LOLE) of the annual peak. Note that scheduled outages are not considered in peak load studies. It is assumed that preventive maintenance will not be scheduled during the period of the year in which the peak load can be expected.

Preventive maintenance of generators contributes significantly to their unavailability. The unavailability consists of two terms: the above-mentioned "forced unavailability" and the "*scheduled unavailability*". The latter is the fraction of time during which a unit is not available due to scheduled outages (i.e. maintenance). The scheduled unavailability of a unit often exceeds its forced unavailability. The scheduled unavailability should not be treated as a stochastic quantity, like the forced unavailability. Generator maintenance can be planned several months or even more than a year ahead. The maintenance planning will be such that the supply of the annual peak load will not be endangered. Typically, maintenance is scheduled away from the annual peak: if the annual peak occurs in winter, generator maintenance is done in summer and the other way around.

A way of including preventive maintenance in the level I evaluation, is to split the year into periods of *e.g.* one week. For each period a LOLE is calculated for the peak load over that period. The generation capacity for each period excludes the units that are in maintenance. Such a study is typically performed as an aid in maintenance scheduling.

The loss-of-load expectation (LOLE) quantifies the risk that the generator capacity is not sufficient to supply the (annual) peak load. It does not quantify the unavailability of the supply due to insufficient generation capacity. To obtain the level I contribution to the unavailability, a more detailed study is required. Not only the unavailability of each generator unit needs to be known, but also its outage frequency and the repair time distribution. The load variation with time and scheduled maintenance has to be taken into account here as well. A simple method is to use the load-duration curve, approximate this through a number of steps and calculate a LOLE for each load level. The application of such calculations is rather limited as they are too complicated to be of use in planning studies, but the influence on the customer is too small to be of any importance. Exceptions are power systems in underdeveloped or very fast developing countries, where lack of generation can seriously contribute to the supply unavailability.

5.2 Transmission Reliability

Level-II reliability concerns the availability of power at so-called bulk supply points: typically transmission substations where power is transformed down to distribution voltage. The power not only has to be generated but also transported to the customers. The availability of sufficient lines or cables has to be taken into account. Level-II reliability studies are much more difficult than Level-I studies, and are still under considerable development. Some of the difficulties are discussed below.

Due to the outage of a transmission line the flow of active and reactive power through the transmission system changes. This can lead to overloading of other lines. The standard example is the overloading of a parallel line. Normally parallel lines will be rated such that the outage of one of them will not lead to overloading of the other. Thus two lines feeding a 200 MVA load should each be able to transport 200 MVA. This so-called *(n-1)* criterion plays an important role in the design of transmission systems: a system consisting of n components should be able to operate with any combination of *(n-1)* components, thus for any single-component outage. In important parts of the system, more strict criteria are used: *(n-2)*, *(n-3), etc.*

The aim of the power system protection is to remove faulted components from the system so as to limit the damage as much as possible. Failure of the protection to remove the faulted component can lead to significantly more damage, including an interruption for customers which would normally not be interrupted. It will be clear that the reliability of the protection is an important part of the reliability of the supply. The power system protection can fail in several ways.

- *The protection fails to operate when required.* In that case the back-up protection will operate and clear the fault. This back-up protection often clears more than only the faulted component making the impact on the system much bigger. Such a protection failure can potentially eliminate the redundancy and lead to an interruption or blackout.

- *The protection operates when not supposed to.* If this happens independently of another event it will simply lead to a outage of the protected component. The redundancy in transmission systems makes that these mal-trips do not have a big influence on the reliability of the supply.

- *The power system protection shows a mal-trip when another relay is supposed to operate.* This leads to the loss of two components at the same time. The large currents flowing through the system during a short circuit make this an event that has to be considered in the calculations. Accurate models for it have not been developed yet. The main problem is that each fault can in theory lead to a mal-trip of any of the other relays in the power system.

- *The power system protection shows a mal-trip due to another event in the system, e.g.* a switching action. Although the event itself does not lead to any required protection intervention, it can still potentially eliminate the redundancy. The reason is that several relays will experience a similar disturbance and thus all might show a mal-trip at the same moment.

The reliability of power system protection is often split into two aspects, called "*dependability*" and "*security*". The dependability is the degree of certainty that the protection will operate correctly (first point); the security is the degree of certainty that the relay will not operate incorrectly (points two through four above). As shown this neglects the different aspects within "security".

Most component outages are due to short circuit faults. Occurrence and clearing of a fault leads to dynamic oscillations in the system. These can lead to overloading or tripping of components. This so-called security aspect of level II reliability is often not taken into account. To include it, detailed dynamic models of the system are needed. In the reliability literature a distinction is made between "*adequacy*" (static evaluation) and "*security*" (dynamic evaluation). The adequacy part is taken care of by most analysis techniques, but security is often left out. In a well-designed transmission system a short circuit should not lead to loss of any generator, or overloading of any component. But especially in heavily loaded systems the dynamic system behavior can have a significant influence on the level-II reliability.

The components in a level-II study are often considered independent, *i.e.* the outage of one component does not depend on the state of the others. But sometimes two or more component outages occur at the same time. Classical examples are the collapse of a tower carrying two circuits and excavation leading to damage of two parallel cables. Several of the other aspects of level-II reliability studies (failure of the protection, overloading of a parallel line) are sometimes also considered common-mode failures. For example: a mal-trip of a relay during a fault on the parallel line will lead to an outage of both lines. By modeling this as common-mode failure, no detailed protection model is needed.

The outage rate is in most studies considered constant, but in reality this is not the case. Many outages are weather-related (lightning, storm, snow) and thus strongly time-dependent. For non-redundant systems this does not matter, but for parallel systems it will significantly increase the interruption rate, even if the average component outage rate stays the same.

5.3 Distribution Reliability

Most published work on power system reliability concerns the generation and transmission systems. Level-III (distribution) reliability studies are rather rare, although this is changing the last few years. The lack of interest in distribution reliability is clearly not due to the high reliability of the distribution system. In fact, both interruption frequency and unavailability as experienced by the customers are mainly determined by events at distribution level, both medium voltage and low voltage.

Distribution systems are most often radially operated. The consequence of this is that each component outage will lead to a supply interruption. To obtain the interruption frequency one only needs to sum the outage rates of all components between the bulk supply point and the customer. Occasionally, parts of the system are operated in parallel or meshed. As this concerns small parts of the system, the mathematical difficulties for calculating the interruption frequency remain limited.

The main problem in distribution system reliability concerns the duration of the interruption. The cost of interruption increases nonlinearly with its duration. The probability distribution function of the interruption duration is of great influence on the expected costs. It is further important to realize that the restoration time depends on the position in the network. The average interruption duration, and thus the interruption costs, can therefore vary significantly throughout the network.

The duration of an interruption consists of a number of terms, each of which has a stochastic character. A well-known law in stochastic theory is that the sum of a sufficient number of stochastic terms has a normal distribution. Thus the distribution of the interruption duration because of its stochastic nature is more likely to be normal than exponential as assumed in most calculation methods. Using the exponential distribution could give unrealistic values for the interruption costs.

6. RELIABILITY AND DEREGULATION

6.1 Is there a conflict?

Deregulation is on one hand seen as potentially leading to reduced reliability. On the other hand deregulation is often mentioned as one of the reasons for the increased interest in reliability. This apparent contradiction can be explained if deregulation is seen as a way of giving a bigger choice to the customers. The customers not only decide about the price of the commodity but also about its reliability, leading to a trade-off between reliability and costs. This trade-off could lead to a lower, equal or higher reliability than in the regulated structure.

Technical constraints make that this free market mechanism is severely limited in electric power systems. As far as the sale of electrical energy is concerned the customer can in a deregulated environment freely choose among different suppliers. But the reliability is to a very large extent determined by the transmission and distribution networks. Both remain natural monopolies, so that the customer cannot freely choose for a less reliable but cheaper supply, or the other way around.

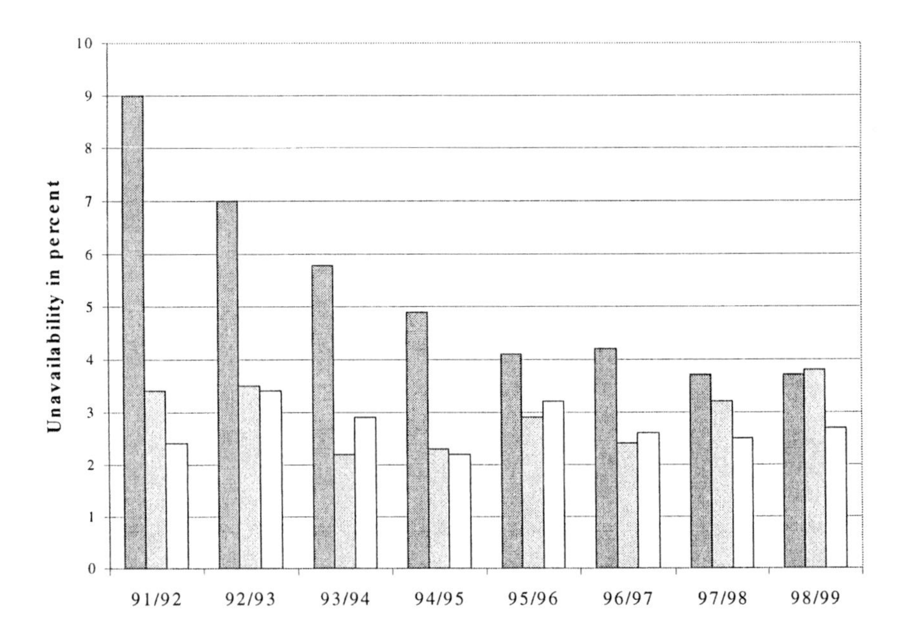

Figure 10. Reduction in transmission unavailability in the United Kingdom

Many customers fear that deregulation and privatization of the electricity industry will lead to less investment, less maintenance, less repair crews, more system overloads due to the increase in long-distance power transports, and to the shutting down of power stations. All this will contribute to a reduction in supply reliability. Customers, especially industrial ones, are concerned that the number and duration of interruptions will increase. Politicians are more concerned about the increasing risk of large-scale blackouts. A number of well-publicized large-scale blackouts in the United States have led to lots of activity within the NERC (North American Electric Reliability Council), the Department of Energy and other bodies, *e.g.* [6].

Consider as an example, *Figure 10*, obtained from reference [13], showing how the unavailability of transmission lines in the UK has dropped since privatization. Data are given for the three transmission operators in the UK: National Grid, Scottish Power and Hydro Electric. One of the operators shows a significant decrease in unavailability, where it stays more or less the same for the other two. At first sight this may be seen as an increase in reliability, until one realizes that only a small part of the unavailability is due to forced outages. The reduction in transmission unavailability is only due to the reduction in the amount of maintenance in the transmission system. Less maintenance is generally accepted as leading to an increased forced outage rate. This may be offset by the large decrease in the scheduled outage rate, but it will almost certainly increase the risk of a large-scale blackout *e.g.* due to protection failure during a short circuit fault. The counter argument here is that the large amount of maintenance done in the past did not contribute to the reliability in the first place. In the deregulated market, maintenance is done more efficient so that the same system reliability is obtained with a smaller scheduled unavailability.

The energy not supplied, shown in *Figure 11*, is obtained from multiplying the duration of interruptions with the amount of load affected by the interruptions. Only interruptions due to outages in the transmission system are included in the data. This is only a small part of the total amount of interruptions, nevertheless it forms an important indicator for the reliability performance of the transmission network. No clear trend is visible in the figure, and the company with reducing amount of maintenance is not worse off than the other two. Note that performance data of transmission network is difficult to interpret, the year-to-year variations are very large due to the low number of events that take place and adverse weather significantly affects the results.

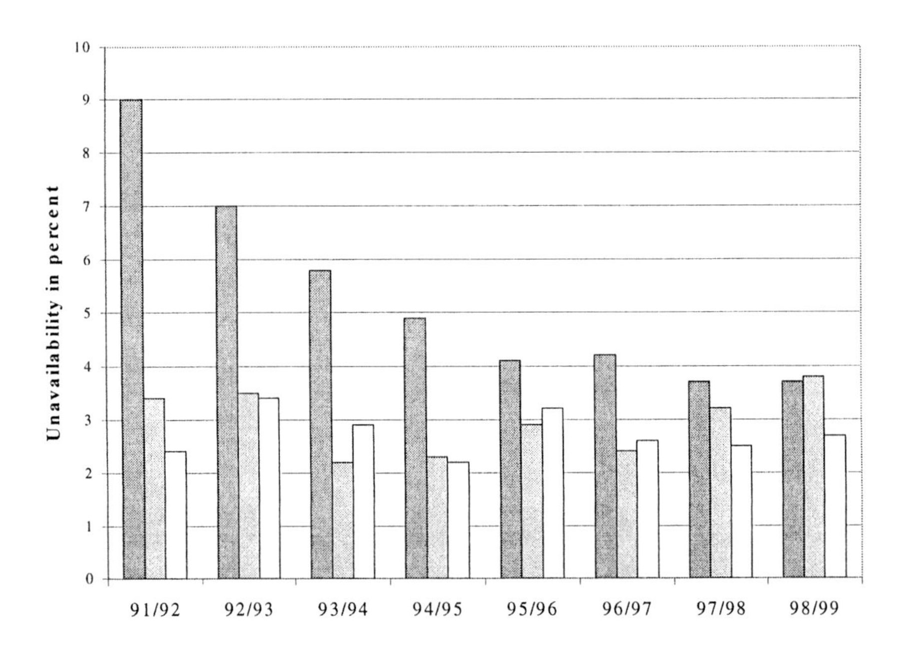

Figure 11. Energy not supplied due to outages in the transmission network in the UK

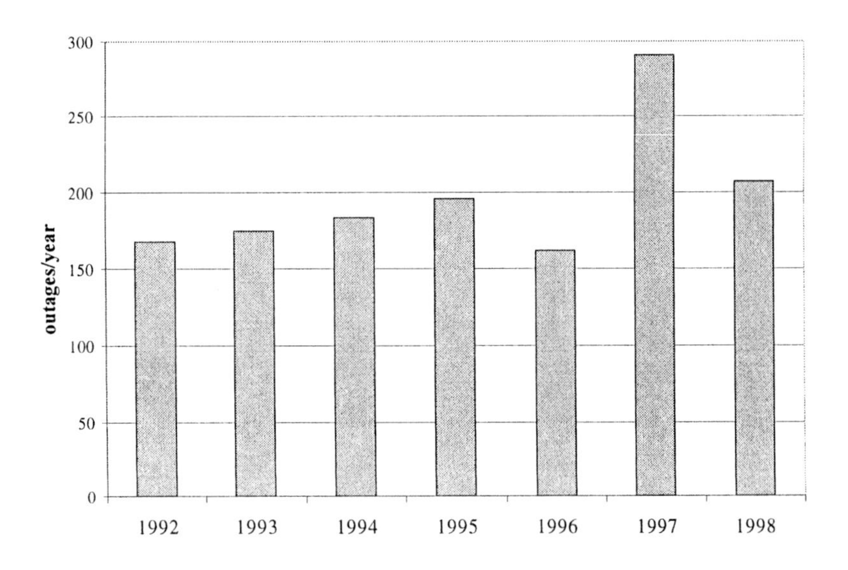

Figure 12. Performance of the Swedish transmission network: number of forced outages

As a comparison, some Swedish data is shown as well. The number of transmission line outages for the Swedish transmission network is shown in *Figure 12* [16]. There is no clear increasing or decreasing trend visible. The amount of un-supplied energy is shown in *Figure 13*; this is what affects the customers. Again it is not possible to draw any hard conclusions from this, but there appears to be an increase in un-supplied energy during the last three years. Whether this is due to deregulation, which started in Sweden in 1996, or due to other (random) effects, is hard to say at this stage.

It is clear to all that deregulation will change the way in which reliability is perceived, by the network operators, by the customers, as well as by politicians. Reliability will no longer be an aim in itself but be subordinate to greater causes (like free market, profit?). Further, it may no longer be obvious that the reliability will be kept at its current value. A reduction in certain aspects of reliability, against a cost reduction of course, could in principle lead to a more desirable situation. The impression one gets from the ongoing discussions however is that the reliability requirements will increase rather than decrease.

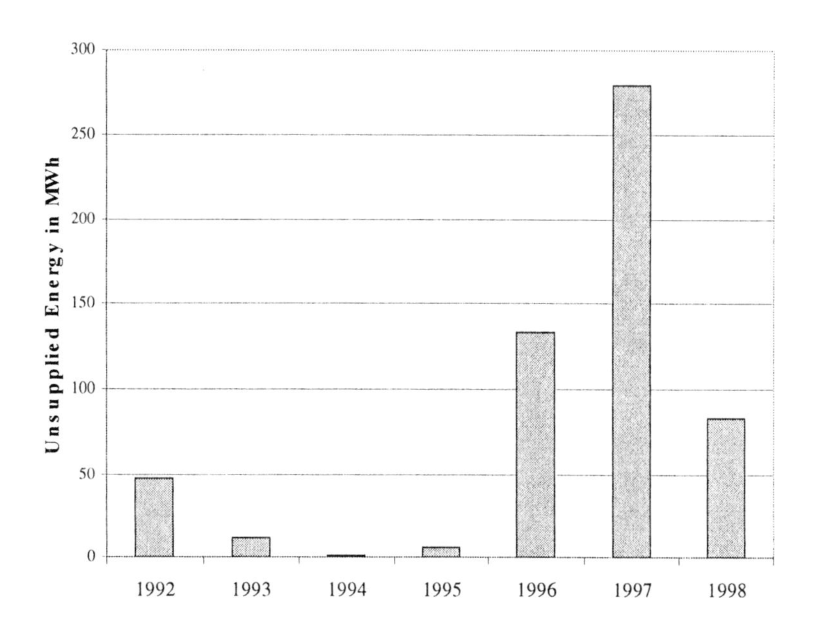

Figure 13. Performance of the Swedish transmission network: Energy not supplied

Not only will the actual reliability of the supply change, also the field of power system reliability analysis is likely to be affected by deregulation. New aspects of reliability may become important, calling for new reliability analysis techniques.

6.2 Reliability analysis

All reliability analysis methods are in their essence planning methods. The power system, or a part of the power system, is modeled and its reliability is calculated for some period in the future. These studies typically look several years ahead, especially studies at transmission and generation level. Distribution reliability may employ a somewhat shorter window, *e.g.* when deciding about the position of net openers in a radial-operated loop system. But even here there is rarely any need for doing the studies on a real-time basis.

In the deregulated power system, changes take place on a day-to-day basis, even on a 15-minute basis in some systems. Neither the load nor the generation is any longer under the control of a central authority, but only following rules of the market. None of the reliability techniques developed in the past is capable of handling changes at such a time-scale. That may look like the end of power system reliability analysis.

Fortunately, the future is not that bleak. The emphasis in power system reliability analysis will obviously have to shift from planning techniques to operational techniques. In fact stochastic prediction techniques are already in use for the operational planning of generator stations. At a given moment in time, for a given generation scheduling and a given load profile, the probability is calculated that the generation will not cover the load at a point in the near future, *e.g.* 15 minutes from now. If this probability is too high, additional units will be started. The further development of such tools will have to become an important aim in the power system reliability field. Universities may play an important role in this. As several recent large-scale blackouts have shown, the risk for the system is not so much in the balance between generation and load but in the transport capacity and the stability of the transmission system. Operation-reliability techniques will have to be developed for transmission systems, which give a warning when the risk of a blackout exceeds a certain threshold.

But even as a planning tool, the end of reliability analysis is not yet in sight. The introduction of the free market merely introduces new levels of uncertainty. As reliability analysis is all about uncertainty, the techniques can still be applied, hence with additional uncertainties included. Again new techniques will have to be developed to handle this.

At generation and transmission level, the main concern is about preventing the risk of a large-scale blackout. Any trade-off between costs and reliability will be very hard to make. Not only are the costs very difficult to estimate, the public disruption due to such an event will be politically unacceptable in most industrialized countries. Interruptions originating at distribution level affect the customer most, not in severity but in number.

The actual interruption costs will be of increasing interest here, so that the duration of interruptions needs to be included much more accurately than can be done by using the exponential distribution. This again calls for new reliability analysis techniques.

6.3 Effects on the Actual Reliability

6.3.1 Generation Reliability

A lack of installed or running generation capacity will lead to an unbalance between production and consumption of electricity. This requires a certain amount of load shedding to prevent an all-out collapse of the power system. In case the unbalance is due to a lack of installed capacity a scheme of rotating interruptions is possible. Care has to be taken with such a scheme that the recovery peak in consumption should not exceed the amount of energy saved during the interruption. Schemes have been proposed in which the customers to be interrupted were determined by a free-market mechanism. In return for a certain incentive customers would "voluntarily" reduce their load or even accept a complete interruption of the supply. For unbalances due to a lack of running capacity an automatic load-shedding scheme is needed, based on under frequency relays. Such schemes are presently in place in most power systems. Typically 10 to 50% percent of customers will be disconnected when the frequency drops below *e.g.* 49 Hz. Market mechanisms could be introduced here as well, where customers that are disconnected first will pay a lower price for their electricity or receive a certain compensation after each disconnection. Interruptions due to unbalances between production and consumption are very rare in most industrialized countries at the moment, although they were common in the fast-growing economies of east Asia only a few years ago. One of the results of deregulation appears to be that the capacity reserve decreases, so that such load shedding schemes may have to be introduced in the future. The Swedish National Grid has during the winter of 1999/2000, allegedly come close to introducing a scheme of rotating interruptions, even though that winter was, by Scandinavian standards, a milder one [21].

A problem with all these schemes is that, it is not possible to point to a responsible party in case of a shortage of generation. The current trends in Europe and elsewhere is to introduce a so-called *independent system operator* (ISO) which is made responsible for reliability. The ISO is often also the operator of the transmission system. It is this ISO that will set up and administer the load-shedding scheme.

Further, the ISO may initiate the building of additional power stations when market mechanism alone will lead to a clear shortage of generation capacity. The current situation in some countries, where a predictable shortage in generation capacity is "solved" by counting on increased import of electricity from abroad, is obviously not a solution in the long run. The problem with intervention is five to ten-year lead-time needed when building large power stations. This requires an early prediction of the future situation or a large safety margin, which could constitute a return to the old regulated system where a high reliability was guaranteed by high investment and thus high electricity prices. Svenska Kraftnät, in its capacity as the Swedish ISO, recently bought six gas turbines with a total power of 400 MW to guarantee the reliability of the supply during peak load situations [19]. It is interesting to note that these were not new gas turbines, but were about to be decommissioned by one of the generation companies. Svenska Kraftnät also called for a reduction of electricity consumption to prevent capacity shortages on extremely cold days, within a few years [17] and it stated explicitly that the responsibility for the supply lays with the generator companies with which customers have signed delivery contracts [18].

There is however a reasoning that predicts, intervention will not be needed in a good functioning market. A potential shortage of electricity due to a lack of generation capacity will lead to an increase in electricity price. This will be an incentive to build more generating stations, as the profit of selling electricity increases. It will also be an incentive to use less electricity. In this way, the market mechanism will influence prices in such as way that the generation capacity is sufficient to cover the demand of electricity. A good example of the operation of the free market is East Asia, where the large shortage of generation capacity in the early 1990's was solved within a few years by opening up the market to independent power producers. Another confirmation of this feedback mechanism is the recent statement by the Svenska Kraftnaä that high electricity prices on the spot market in anticipation of a high demand actually led to a lower demand than expected [20].

Such feedback mechanisms work when there is no price-cap on the electricity price. The introduction of a price-cap means that generators cannot increase prices to such values that could influence consumption.

6.3.2 Transmission Networks

Transmission system reliability is associated with the limited transport capacity of any transmission network. A lack of transport capacity will lead to a so-called network overload, and can lead to voltage collapse and large-scale blackouts. System overload is used here in a very broad sense as any situation in which a large-scale blackout can be prevented by reduction of the loading of transmission lines. Even the total loss of power to a transmission substation is considered an overload situation here: one with a transport capacity zero. In reality there are very different mechanisms behind the various phenomena, each deserving a more in-depth treatment than can be offered here.

When considering mitigation methods for network overload it is essential to distinguish between predictable and non-predictable overloads. To mitigate a *predictable network overload*, a load-shedding scheme can be set up, similar to the one used to prevent shortages of generation capacity discussed before. An important difference is that for a network overload no longer all customers are equal. Load shedding on the receiving end of an overloaded line will have different effects than on the sending end. In fact, the latter is likely to have no effect at all on the overload. As only part of the load can be considered for load shedding, market mechanisms alone may not be enough for this. Some additional intervention by the ISO is required. The ISO could decide that load shedding is necessary, but market incentives could decide which customers are disconnected first. A serious technical problem with these load-shedding schemes is that often a very fast decision has to be made. With production shortages the frequency of the system can be used as an indicator. With system overload no such global indicator exists. Extensive research has been done towards finding indicators for the proximity to voltage collapse, but it is not possible to make such a decision based on measurements done at a certain point in the system. Any decision will have to be made centrally and this decision has to be communicated to all customers involved.

Non-predictable network overloads are typically due to the forced outage of a transmission line or of a power station. A load-shedding scheme that only comes in action after the outage is all but impossible in such a case. The risk of a non-predictable network overload could be reduced through preventive load shedding. When the system is operated close to its limit, when the $(n-1)$-criterion is no longer upheld, some load could be removed or a scheme of rotating interruptions could be introduced. It may however be very difficult to justify such scheduled interruptions to customers, unless they are accompanied by high financial advantages. Obviously a completely free market would also here be possible, but not necessarily desirable.

Large-scale blackouts due to (predictable or non-predictable) transmission overload are again very rare. The increase in large-scale power transport is one of the consequences of deregulation. According to [9] the volume of transactions has increased by about 50% per year for the last several years. These increased power flows will almost certainly lead to an increase in the risk of blackouts. The proper introduction of market forces may also here significantly reduce the risk, but not enough is known about these mechanisms yet to assess how large the risk will be.

Newly introduced power-electronics based series and shunt controllers (Flexible AC Transmission Systems or FACTS) have the potential to significantly increase the capacity of the system, so that overloads can be mitigated without any reduction in load. In the long term however, overload problems may still occur but at a (much) higher loading level. Despite these kind of technical solutions, each network will always have a maximum transport capacity. Furthermore, these devices may make the system very vulnerable where a single failure in a power-electronics controller triggers a large-scale blackout.

A short-circuit in the transmission system may cause transient instability leading to the loss of power stations and large-scale blackouts. Large power transfers through the transmission system increase the risk of transient instability and thus of blackouts. The main difference with the two above phenomena is that transient stability is a very fast phenomenon. It will within a few seconds result in a very severe unbalance between production and consumption as well as the system being split up in various islands, some of which may "survive", others not. The responsibility for transient instability lays in principle again with an ISO. Preventing it requires limitations of power transfer, enforcement of the transmission system, or faster fault clearing. The first will hinder the free market, the second will meet severe environmental constraints, so that the latter may be the solution to go for. This obviously requires investment in the network, thus leading to an increase in operation costs, to be shared by all customers. How far this reduction in fault-clearing time should go, remains an open question. A permanent fault requires the complete removal of the line from service, which in turn can lead to overload problems like voltage collapse.

Despite all the market mechanisms mentioned above to reduce load during emergencies, blackouts cannot be completely prevented. A blackout is here defined as a forced interruption of a large group of customers over a wide area. Voluntary or market-controlled interruptions are not considered as blackouts. As the transmission system remains a natural monopoly, there are not many incentives for the system owner to prevent these forced blackouts. The combination of ISO and transmission system operator in one company does not make this more transparent.

A scheme should be introduced in which an independent body assesses the system reliability and forces the network operator to improve reliability where needed. The North American Electric Reliability Council (NERC) is setting up a structure to guarantee the reliability of generation and transmission. A serious problem will all such schemes is that observing the reliability (i.e. counting blackouts) will only lead to statistically relevant results after a long monitoring period. Also the aim should be to prevent blackouts before they actually occur. Unfortunately power system reliability techniques for transmission systems have not yet proven their accuracy either.

6.3.3 Distribution Networks

The majority of interruptions experienced by a customer is due to short-circuit faults in the distribution network leading to the removal of the faulted feeder by the power system protection. Costs reduction measures at distribution companies may lead to a reduced number of maintenance and repair crews. Reduced maintenance will lead to an increased number of faults where less repair crews makes that the duration of an interruptions is likely to be longer. Counter arguments against this are that maintenance could be done more efficient by less personnel *e.g.* by introducing reliability-centered maintenance. Also the introduction of distribution automation will lead to faster restoration of the supply while at the same time reducing the personnel costs.

In any case it appears appropriate to force distribution companies to publish their performance. A very good scheme for this has been implemented in the UK. Each year a report is published by the Office of Electricity Regulation (OFFER) in which the number of interruptions and the supply unavailability are given for each distribution company. Similar information is published for the transmission system, [11, 13]. Also in many other countries transmission and distribution companies keep track of their reliability performance, so that such a scheme could be easily introduced elsewhere. Unfortunately the information is rarely made public in such a way as in the UK.

6.4 Regulation of the Market

6.4.1 Responsibility

In a deregulated market, the responsibility for the supply reliability becomes spread over several companies and organizations. This makes it hard to define the responsibility for a specific event. For example: the outage of a transmission line due to a fault caused by a third party, triggers a blackout, but this blackout could have been prevented if a generating station had not been out of operation for maintenance. Who is to blame for this blackout: the third party causing the fault; the transmission network operator that didn't operate with enough redundancy; the generator operator that decided to schedule maintenance while the network was without sufficient redundancy; or the contractor that performed the maintenance? It is obvious that nobody can be made responsible for such an individual case without very strict rules on reliable operation of transmission systems. The (*n-1*)-criterion could be a useful basis for a legal framework defining the responsibility of the different partners. As far as generation and transmission reliability is concerned, the ISO needs to play an important role in setting up these rules. The ISO should have the authority to intervene in the market to guarantee the reliability of the system.

At distribution level the situation is somewhat easier to define. The system is radially operated so that each outage will lead to an interruption. The restoration of the supply, and thus the duration of the interruption, are also determined by the operation and loading of the rest of the system. But in almost all cases, the responsibility can be placed with the distribution network operator.

6.4.2 Contracts and Compensation Schemes

In Europe, the base line of reliability and voltage quality is defined in the voltage characteristic standards EN 50160. On long interruptions, the event considered in reliability discussions, the document states that "*under normal operating conditions the annual frequency of voltage interruptions longer than three minutes may be less than 10 or up to 50 depending on the area*". The document also states that "*it is not possible to indicate typical values for the annual frequency and duration of long interruptions*". Even when we neglect the clause on normal operating conditions, the document does not constitute a serious reliability requirement or guarantee.

Many customers want more accurate limits for the interruption frequency. Therefore some network operators offer their customers special guarantees, sometimes called "power quality contracts". The network operator sets a limit to the unavailability and/or to the number of interruptions per year. If this limit is exceeded in a given year, the utility pays compensation to the customer. This can be an amount as defined in the contract, or the actual financial loss to the customer. Some utilities offer various levels of quality, with different costs. The number of options is almost unlimited: customer willingness to pay extra for higher reliability and utility creativity are the main influencing factors at the moment. Technical considerations do not appear to play any role in setting levels for the maximum number of interruptions or the costs of the various options. For a customer to make a decision about the best option, data should be available, not only about the average interruption frequency but also on the probability distribution of the number of interruptions per year.

A special case is where the local utility owns the network as well as generation, a typical US situation. The utility may decide to offer additional reliability guarantees to its local customers in return for long-term contracts or slightly higher prices. Other generating companies can still include compensation schemes in their contracts, but they will not be able to affect the supply reliability. Complete separation of the network and power generation, as is the current trend in Europe, will make such constructions are no longer possible.

Another alternative is where the transmission or distribution network operator pays compensation to all affected customers, thus as part of the standard contractual obligations. Such a scheme could be set up voluntarily by the company, but more likely the scheme is enforced by the government, or by an independent regulator. Distribution companies in the UK have to offer a fixed amount to each customer interrupted for longer than 24 hours. In the Netherlands a court ruling states that network operators have to compensate domestic customers for all interruption costs exceeding a certain amount. However, if the blame lays with a third party, the utility is not liable to pay compensation (neither is the third party according to the current interpretation of the court ruling). Also in Sweden some utilities offer customers compensation for an interruption.

None of these schemes actually guarantees a certain reliability, it only guarantees compensation in case the reliability gets too low. These schemes are merely an insurance against supply interruptions, which can equally well be offered by an insurance company outside of the electricity market. A disadvantage of all these schemes is that network operators may be tempted to postpone investments if that would save more than the costs of paying compensation. The role of a regulator, or an independent system operator, is to force the transmission and distribution companies to maintain a certain level of reliability. The introduction of design standards is part of the solution, but next to that the recording of the reliability performance remains important. Keeping track of the performance of transmission and distribution companies enables a regulator to intervene if the reliability threatens to go down too much. This holds especially for distribution companies where the number of interruptions is big enough to give confident statistical results. For transmission companies some kind of predictive indices may have to be introduced, in combination with performance statistics.

6.4.3 Performance standards

The inconvenience of an interruption increases very fast when its duration exceeds a few hours. This holds especially for domestic customers. Therefore it makes sense to reduce the duration of interruptions instead of their frequency (which could be very expensive). Limiting the duration of interruptions is a basic philosophy in power system design and operation in almost any country. In the UK, as an example, the duration of interruptions is limited in three ways:

OFFER sets targets for the percentage of interruptions lasting longer than 3 hours and for the percentage of interruptions lasting longer than 24 hours. These are so-called "*guaranteed standards of service*". Some are [11]:

- The electricity supply after a fault must be restored within 24 hours. The penalty for not meeting this standard is £50 for domestic customers and £100 for non-domestic customers plus £25 for each subsequent 12 hour. The penalties were increased in 1998, and the latest distribution price-review proposes to bring back the maximum duration of an interruption to 18 hours [10].
- The distribution company should give a customer at least 5 days notice of planned supply interruptions. Penalties are £20 for domestic customers and £40 for non-domestic customers.
- A customer complaint regarding voltage, should be followed by a visit within 7 working days or a substantive reply within 5 working days. The penalty is £20.
- Distribution companies should offer and keep morning or afternoon appointments or a timed appointment if requested by the customer. The penalty is £20.

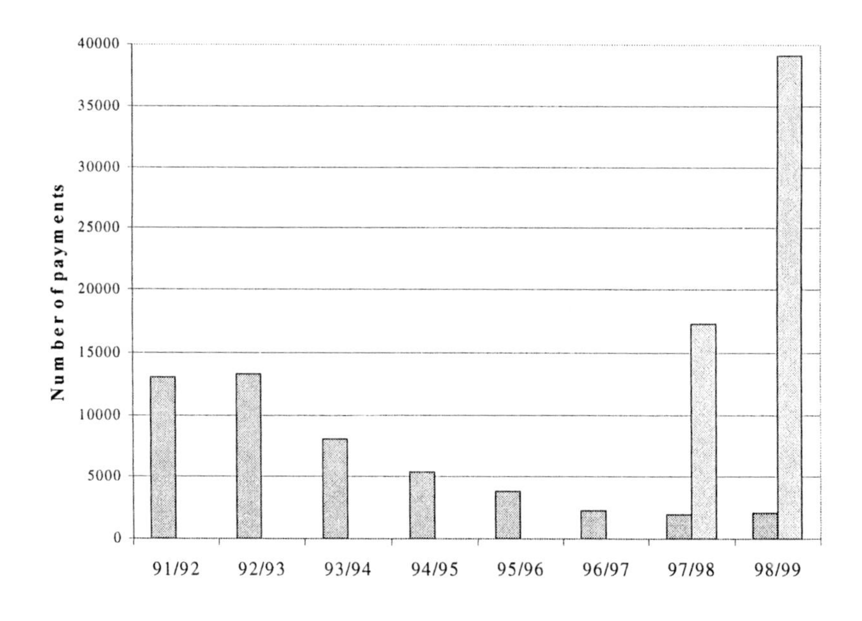

Figure 14. Number of payments under the guaranteed standards scheme in the UK

The number of payments under the guaranteed standards scheme is published annually, subdivided for each standard and for each distribution company. Everybody is able to compare the performance of the different distribution companies. The number of payments since privatization is shown in *Figure 14* [11, 12]. Note the two values for 97/98 and 98/99. The first ones exclude interruptions during winter storms. The second one covers all interruptions. During the last two years, interruptions during winter storms are no longer exempt from the guaranteed standards scheme. Neglecting the winter storms, the number of payments shows a clearly-downward trend, indicating an improvement in reliability.

Bad weather periods in the UK are mainly in the form of winter storms. One of these storms hit Scotland and the northern part of England on Boxing Day 1998 [14]. Many customers were interrupted for longer than 24 hours. This storm is responsible for the 98/99 peak in *Figure 14*. Some utilities decided to pay voluntary compensation, sometimes even higher than the compensation under the guaranteed standards scheme. By paying voluntary compensation, these companies would not see their statistics negatively affected by the storm. The total interruption frequency and unavailability would still be affected, but the influence of one storm, even such a severe one, on those statistics is much more limited. The number of payments and the amount of compensation paid are summarized in *Table 1* [12].

Table 1. Winter storm payments 1998/99

Company	Guaranteed standard payments	Ex-gratia payments	Total amount paid
A	14 180	47	£ 511 660
B	22 846	420	£ 934 276
C	-	5	£ 164
D	-	51	£ 2 650
E	-	533	£ 26 875
F	-	50 990	£ 4 506 708
G	-	583	£ 47 000

Next to these "guaranteed standards" a number of "*overall standards of service*" are in place in the UK [12]. An overall standard is a statistic that a distribution company must keep, but there are no direct penalties for not keeping them. In the long term a company's performance with respect to overall standards is considered in the setting of a ceiling for the electricity price which a distribution is allowed to charge. Some examples of overall standards are:

- A minimum percent of interruptions must be restored within three hours.
- A minimum percent of interruptions must be restored within 24 hours.
- A minimum percent of under-voltages or over-voltages must be corrected within six months.
- A minimum percent of all customer letters must be responded to within 10 working days.

Each overall standard is associated with a minimum percentage. This value may be different for each distribution company, depending on specific local conditions. Percentages may also be changed, typically increased, from year to year. For most overall standards the percentage has reached a value of 100%. A clear exception to this is the first standard where values between 80% and 95% are in place. The performance of the distribution companies is again published annually. The target and the performance of the first overall standard is shown in *Figure 15* for two distribution companies. Note that the target has been increased for both companies when it turned out that their performance well exceeded the target. An increase in target can be initiated by the regulator or by the distribution company. If a company managed to keep a high target, this will positively effect the ceiling on the electricity price.

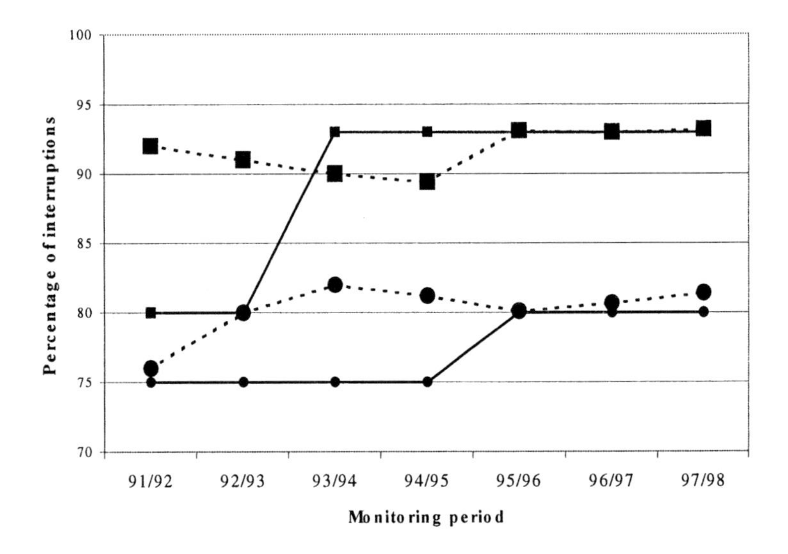

Figure 15. Target (solid lines) versus performance (dashed lines) for the percentage of interruptions restored within three hours, for a rural (squares) and an urban (circles) distribution company in the UK.

7. PERFORMANCE INDICATORS

When comparing the performance of different distribution and transmission companies, or tracking the performance of one company over time, it is important to clearly define reliability performance indicators. Deregulation does not require new definitions here, as interruptions have been tracked by utilities for many years already. The two main performance indicators are the interruption frequency and the supply unavailability.

Traditionally only forced interruptions are considered, although again in the U.K. also statistics on scheduled interruptions are published. In the future statistics for a third category, "market-driven interruptions" may have to be introduced. Below only indicators for forced interruptions are presented.

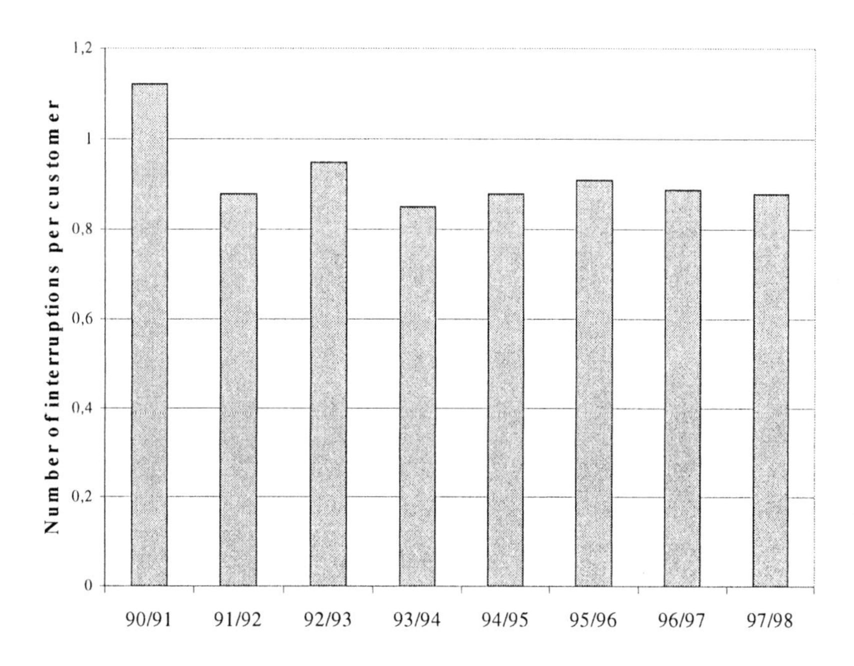

Figure 16. Number of supply interruptions per customer, UK data

The interruption frequency λ (or average number of interruptions per customer per year) is defined as:

$$\lambda = \frac{\sum_{i=1}^{K} N_i}{N_{tot}} \tag{25}$$

N_i is the number of customers affected by interruption i, N_{tot} the total number of customers fed from the system, and k the number of interruptions during the observation period. The supply unavailability q is defined as:

$$q = \frac{\sum_{i=1}^{k} N_i D_i}{N_{tot}} \tag{26}$$

D_i is the duration of interruption i. The unavailability is typically expressed in minutes per year. From interruption frequency and unavailability the average duration of an interruption can be obtained as follows:

$$\overline{D} = \frac{q}{\lambda} \tag{27}$$

The annual interruption frequency for the UK is given in *Figure 16* for the period since privatization. The value is rather constant, despite a variation in weather from year to year. Also the performance for each of the distribution companies is rather constant. The corresponding supply unavailability is shown in *Figure 17*. The large peak for 90/91 was due to a snowstorm hitting the central part of the UK. Note that this effected the unavailability much more than the interruption frequency. Note also that the before-mentioned boxing-day storm did not significantly affect the performance over 97/98.

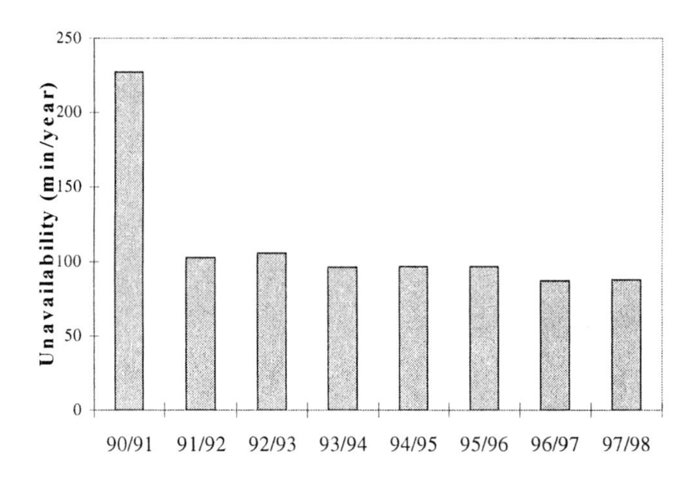

Figure 17. Supply unavailability for UK customers

8. CONCLUSIONS

The natural monopoly of the transmission and distribution network operators has created a well-founded concern that the reliability of the power supply will decrease in the future. To prevent this, a certain amount of regulation is unavoidable. A regulatory body needs to collect and publish performance indicators of all transmission and distribution companies. The regulatory body also needs to set standards for the reliability performance, including a penalty system. Possible options are the payment of compensation to customers, and a control of the distribution and transmission charges depending on the reliability performance. The techniques for data collection and the calculation of performance indicators is well developed and can directly be used.

The risk of a serious blackout requires a somewhat different approach. The consequences of such an event are so large that it is not appropriate to wait for the collection of sufficiently confident statistics. Some kind of stochastic prediction of the risk of a blackout needs to be applied to the system. When this risk becomes unacceptably high, the regulatory body should intervene. In many deregulated markets, this task lays with the operator of the transmission grid. Unfortunately, implementation of the task is not very transparent. The main problems are expected at transmission level. The free market at generation level may prevent large shortages in generation capacity.

Various market mechanisms may also be used to prevent too much reduction in supply reliability, especially to prevent shortages in generation capacity and in transport capacity in the transmission network. New reliability analysis tools need to be developed to include the uncertainties of the market in reliability planning tools, and to enable the application of reliability techniques at the system operational level.

Despite the fear for a decreased reliability, there are no indications at the moment that reliability is indeed reducing. Neither interruption frequency, nor unavailability shown an upward trend. Some recent blackouts and almost near blackout situations did however point to the vulnerability of the system.

REFERENCES

1. R. Billinton and W. Li, *Reliability assessment of electric power systems using Monte Carlo simulation*, New York: Plenum, 1994.
2. M. H. J. Bollen, "Method for reliability analysis of industrial distribution systems", IEE Proceedings-C, Vol.140, no.6, Nov. 1993, p.497-502.
3. M. H. J. Bollen and P. E. Dirix, "Simple motor models for reliability and power quality assessment of industrial power systems", IEE Proceedings– Generation, Transmission, Distribution, January 1996, pp.56-60.
4. J. van Casteren and M. H. J. Bollen, "Fast analytical reliability worth assessment using non-homogeneous models. Probability methods applied to electric power systems" (PMAPS), September 2000, Madeira, Portugal.

5. J. van Casteren, M. H. J. Bollen and M. Schmieg, "Reliability analysis of electrical power systems: The Weibull-Markov model", IEEE Transactions on Industry Applications, in print.
6. Interim report of the US Department of Energy's power outage study team, Findings from the summer of 1999.
7. H.J. Koglin, E. Roos, H. J. Richter, U. Scherer and W. Wellssow, "Experience in the reliability evaluation of high voltage networks", Proc. of International Conference on Large High Voltage Electric Networks (CIGRE), September 1986, Paris, France.
8. D. O. Koval and J. P. Ratusz, "Substation reliability simulation model", IEEE Transactions on Industry Applications, Vol.29, September/October 1993, pp.1012-1017.
9. W. K. Newman, "Transmission structure issues, in: Reliability in the new market structure", IEEE Power Engineering Review, December 1999, pp.5-6.
10. The distribution price control review and separation of businesses, Office of Gas and Electricity Markets, Birmingham, UK, 8 October 1999.
11. Report on customer services 1997/98, Office of Electricity Regulation, Birmingham, UK, 1998.
12. Report on customer services 1998/99, Office of Gas and Electricity Markets, Birmingham, UK, 1999.
13. Report on distribution and transmission system performance 1997/98, Office of Electricity Regulation, Birmingham, UK, 1998.
14. Supply interruptions following the Boxing Day storms 1998, Office of Electricity Regulation, Birmingham, UK, May 1999.
15. B. H. T. Smeets and M. H. J. Bollen, "Stochastic modeling of protection systems– comparison of four mathematical techniques", Eindhoven University of Technology, Eindhoven, The Netherlands, June 1995. Research report 95-E-291.
16. Svenska Kraftnät, Annual Report 1998.
17. Svenska Kraftnät, press release, 5 October 1999.
18. Svenska Kraftnät, press release, 8 October 1999.
19. Svenska Kraftnät, press release, 3 January 2000.
20. Svenska Kraftnät, press release, 24 January 2000.
21. Svenska Kraftnät, personal communication.

FURTHER READING

1. R. Allan and R. Billinton, "Probabilistic assessment of power systems", Proceedings of IEEE, Vol.88, February 2000, pp.140-162.
2. R. N. Allan, R. Billinton, A. M. Briepohl and C. H. Grigg, "Bibliography on the application of probability methods in power system reliability evaluation 1987-1991", IEEE Transactions on Power Systems, Feb. 1994, pp.41-49.
3. R. N. Allan, R. Billinton and S. H. Lee, "Bibliography on the application of probability methods in power system reliability evaluation, 1977-1982", IEEE Transactions on Power Apparatus and Systems, 1984, pp.275-282.
4. R. N. Allan, R. Billinton, S. M. Shadidehpour and C. Singh, "Bibliography on the application of probability methods in power system reliability evaluation, 1982-1987", IEEE Transactions on Power Systems, 1988, pp.1555-1564.
5. R. N. Allan, R. Billinton, A. M. Briepohl and C. H. Grigg, "Bibliography on the application of probability methods in power system reliability evaluation, 1987-1991", IEEE Transactions on Power Systems, 1994, pp.41-49.

6. R. N. Allan, R. Billinton, A. M. Briepohl and C. H. Grigg, "Bibliography on the application of probability methods in power system reliability evaluation, 1991-1996", IEEE Transactions on Power Systems, 1999, pp.51-57.
7. R. Billinton, R. N. Allan and L. Salvaderi (Edited), *Applied reliability assessment in electric power systems*, New York: IEEE Press, 1991.
8. R. Billinton, "Bibliography on the application of probability methods in power system reliability evaluation", IEEE Transactions on Power Apparatus and Systems, 1972, pp.649-660.
9. R. Billinton and R. N. Allan, *Reliability assessment of large electric power systems*, Kluwer Academic Publishers, 1988.
10. R. Billinton and R. N. Allan, *Reliability evaluation of power systems*, Second Edition. New York: Plenum, 1996.
11. M. H. J. Bollen, *Understanding power quality: voltage sags and interruptions*, New York: IEEE Press, 1999.
12. M. H. J. Bollen, "Literature search for reliability data of components in electric distribution networks". Eindhoven University of Technology Research Report 93-E-276, Aug. 1993.
13. S. Burns and G. Gross, "Value of service reliability", IEEE Transactions on Power Systems, August 1990, pp.825-834.
14. J. Endreyni, *Reliability modeling in electric power systems*, Wiley, 1978.
15. N. Hingorany and Guygi, *Understanding FACTS*, New York: IEEE Press, 2000.
16. IEEE Recommended Practice for the Design of Reliable Industrial and Commercial Power Systems (The Gold Book), IEEE Std.493-1997.
17. IEEE Subcommittee Report, "Bibliography on the application of probability methods in power system reliability evaluation, 1971-1977", IEEE Transactions on Power Apparatus and Systems, 1978, pp.2235-2242.
18. K. K. Kariuki and R. N. Allan, "Assessment of customer outage costs due to service interruptions: residential sector", Proc. IEE: Generation, Transmission, Distribution, Vol.143, 1996, pp.171-180.
19. K. K. Kariuki and R. N. Allan, "Application of customer outage costs in system planning, design and operation", Proc. IEE: Generation, Transmission, Distribution, Vol.143, 1996, pp.305-312.
20. K. K. Kariuki and R. N. Allan, "Factors affecting customer outage costs due to electric service interruptions", Proc. IEE: Generation, Transmission, Distribution, Vol.143, 1996, pp.521-528.
21. H. D. Kochs, *Zuverlässigkeit elektrotechnische anlagen,* Berlin, Germany: Springer, 1984.
22. A. Mäkinen, J. Partanen and E. Lakervi, "A practical approach to estimating future outage costs in power distribution networks", IEEE Transactions on Power Delivery, January 1990, pp.313-316.
23. A. M. Shaalan, "Electric service interruptions: impacts and costs estimation", Electra (CIGRE), no.127, 1989, pp.89-109.
24. G. Wacker, E. Wojczynski and R. Billinton, "Interruption cost methodology and results– A Canadian residential survey", IEEE Transactions on Power Apparatus and Systems, October 1983, pp.3385-3392.
25. G. Wacker and R. Billinton, "Customer costs of electric service interruptions", Proceedings of the IEEE, Vol.77, no.6, June 1989, p.919-930.
26. G. Wacker and R. Billinton, "Customer costs of electric service interruptions", Proceedings of the IEEE, Vol.77, no.6, June 1989, pp.919-930.
27. http://www.nerc.com North American Electric Reliability Council

28. http://www.svk.se Svenska Kraftnät, the Swedish ISO
29. http://www.ofgem.gov.uk Office of Gas and Electricity Markets for the UK
30. http://www.detroitedison.com Detroit Edison, Michigan, US.

Chapter 7

Power Quality Issues

Voltage Dips and other Disturbances

1. POWER QUALITY

Power quality has become a commonly used term in discussions on performance of the power supply. Like reliability, there is a serious concern that the power quality will decrease due to deregulation. In this chapter, the most important issues in power quality will be discussed. Both voltage dips and short interruptions are a concern for many domestic, industrial and commercial customers. The effects of an individual event are more severe for long interruptions than for short interruptions or voltage dips. But due to the higher frequency of occurrence, the latter two is of much higher concern.

A fundamental difference between voltage dips and short interruptions is in their origins. Short interruptions always originate in the local distribution network, close to the affected customers. This makes mitigation methods and the question of responsibility rather easy, albeit not always cheap. Faults anywhere in the system, even on transmission lines hundreds of kilometers away, can lead to a serious voltage dip. This makes mitigation much more difficult. Also the responsibility questions becomes very complicated, as will be shown later.

1.1 Terminology

The term "power quality" is used in several different contexts. An overview on such confusion, is clearly outside the scope of this chapter, but certainly makes interesting reading. A consistent set of definitions, giving an insight in the various aspects of power quality [1] is as follows:

- *Voltage quality* concerns the deviation of the voltage waveform from the ideal. The ideal is a sinusoidal voltage of constant magnitude and constant frequency. This term only covers the technical aspects and does not cover any non-idealities in the current. Voltage quality involves the performance of the power system towards the load.

- *Current quality* is a complimentary term of voltage quality. It concerns the deviation of the current waveform from the ideal. The ideal current is again sinusoidal with constant magnitude and frequency, but also in phase with the voltage waveform. Current quality involves the performance of the load towards the power system.

- *Power quality* is the combination of current quality and voltage quality: involving the interaction between the system and the load. Note that we use the terms "system" and "load", and not "utility" and "customer". The former are technical terms, the latter are "legal" terms. Power quality, as defined here, is a technical issue. Obviously, since it has to do with performance, many legal consequences do follow the technical issues.

- *Quality of supply* covers the technical issues involved in voltage quality, plus all non-technical issues in the performance of the utility towards the customer.

- *Quality of consumption* is a complementary term to quality of supply, covering current quality plus the performance of the customer towards the utility.

In the IEC standards, the term electromagnetic compatibility is used, abbreviated often as EMC. Electromagnetic compatibility is defined as "the ability of a device to operate correctly in its electromagnetic environment without unduly polluting the electromagnetic environment" [12]. From various standard documents it becomes clear that the electromagnetic environment includes voltage quality as defined above. Issues such as light flicker due to voltage fluctuations and harmonic distortion are treated in a number of IEC standards. The American standards setting organization IEEE does use the term "power quality" and has published a number of documents on this subject. Although strictly speaking only applicable to the US, they do contain information of interest to non-US readers too. A number of these documents even give textbook style overview of certain power quality aspects. A list of relevant IEC and IEEE standards on power quality is given in Appendices A and B with this chapter. These overviews give a good impression of the width of this area, only a very small part of which is treated in this chapter.

1.2 Interest in Power Quality

Several reasons have been given for the recent interest in power quality issues. The introduction of new types of end-user equipment is generally cited as the drive behind the increased activity in this area, but that alone cannot explain everything. The following list of reasons is probably rather complete:

- *Equipment has become less tolerant to voltage disturbances:* Not only will modern equipment trip much faster, but also production processes and companies have become less tolerant to equipment trips.

- *Equipment causes voltage disturbances:* Often the same equipment that is intolerant to voltage disturbance will itself cause other voltage disturbances.

- *The need for performance criteria:* Partly due to deregulation, there is an increased need for performance criteria to assess how good the power companies do their job. The step from performance criteria to power quality is only small.

- *Utilities want to deliver a good product:* Many power quality developments are driven by utilities. Most utilities simply want to deliver a good product and have been committed to that for decades.

- *The power supply has become too good:* In most industrial countries long interruptions and blackouts have become a rare phenomenon. The result is an increased attention for "minor problems" such as voltage dips and short interruptions. Expressed more positively, this can be seen as part of the continuous drive towards better engineering systems.

- *Power quality can be measured:* The availability of cheap power quality monitors means that, for the first time, voltage and current quality can actually be monitored on a large scale.

1.3 Events and Variations

Power quality phenomena are deviations from the ideal voltage and current. An overview of the phenomena is given in several of the publications on the further reading list. The phenomena are of two principally different types. The so-called "*variations*" are deviations that are always present, such as the (small) difference between the nominal and the actual system voltage. The other class: "*events*" are the occasional (larger) deviations such as voltage dips and interruptions.

The majority of voltage events of interest to customers involve either an increase or a decrease of the voltage magnitude. A *voltage magnitude event* is defined as a (significant) sudden deviation from the normal voltage magnitude for a limited duration. The (magnitude of) remaining voltage and the duration of the event are the main characteristics of each event. Different names are in use for events with different characteristics. An overview of the terminology in use is given in *Figure 1.* Note that the event magnitude is defined here as the remaining voltage, whereas some publications and the existing IEC documents actually use the voltage drop or rise to characterize the event.

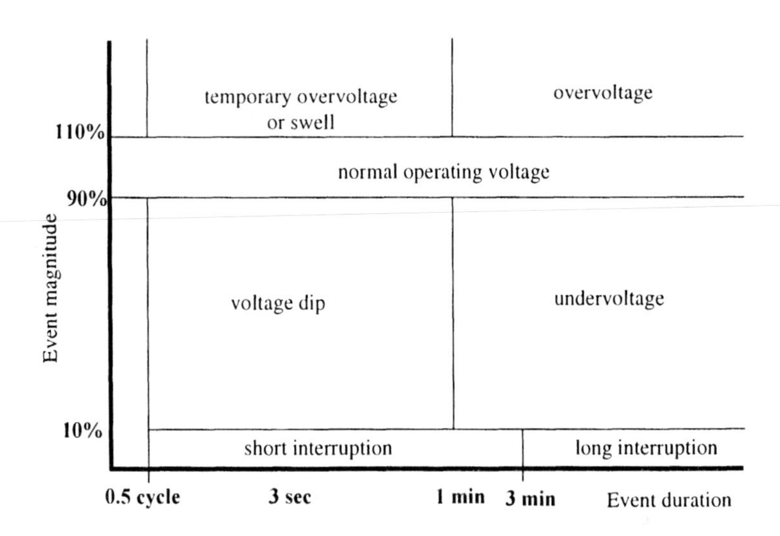

Figure 1. Classification of voltage magnitude events

An alternative classification of voltage magnitude events is proposed in [1], where distinction is made in event duration between:

- *very short* (up to 1-3 cycles), corresponding to transient and self-restoring events;
- *short* (1-3 cycles through 1-3 minutes), corresponding to automatic restoration of the pre-event situation;
- *long* (1-3 minutes through 1-3 hours), corresponding to manual restoration through switching actions;
- *very long* (longer than 1-3 hours), corresponding to repair or replacement of the faulted component.

2. VOLTAGE DIPS

2.1 Magnitude and Duration

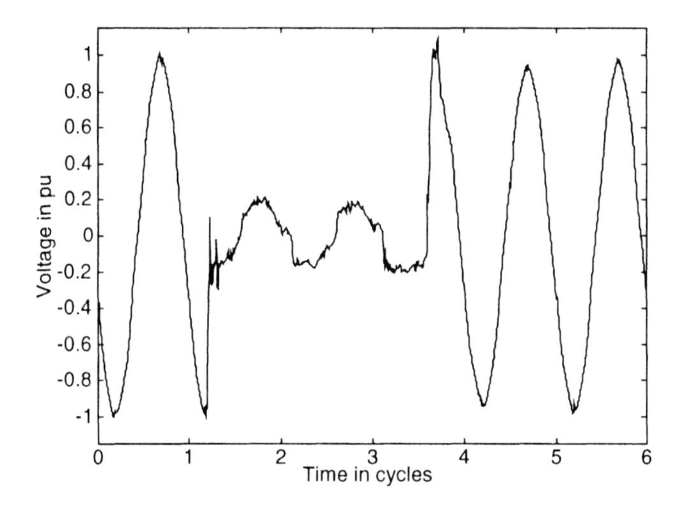

Figure 2. Example of a voltage dip: instantaneous voltage versus time.

Voltage dips are short duration reductions in rms voltage, mainly due to faults and starting of large motors. The large interest in voltage dips is due to the problems they cause for several types of equipment essential in commercial and industrial installations. Especially computing equipment, adjustable-speed drives and process-control equipment are notorious for their sensitivity. Some pieces of equipment trip when the rms voltage drops below 90% for just one or two cycles. For dips down to 70% during 100 ms, a large fraction of drives and electronic equipment will mal-operate.

An example of a voltage dip is shown in *Figure 2*: the plot shows the instantaneous voltage versus time as obtained from a power quality monitor. The monitor triggers when the magnitude of the voltage drops below a pre-set threshold or when the instantaneous voltage differs more than a pre-set amount from the instantaneous voltage one cycle back in time. Using a cyclical memory the monitor captures a certain number of samples pre-trigger. The recording in *Figure 2* shows one cycle of pre-dip voltage. Suddenly the voltage drops to a lower (rms) value, where it stays for about two cycles after which the voltage recovers again. Several more examples of voltage dips will be shown in the remainder of this chapter.

The magnitude and duration of a voltage dip are determined from the rms voltage as a function of time. The rms voltage is typically determined over a one-cycle window:

$$V_{rms}(k) = \sqrt{\frac{1}{N} \sum_{i=k-N+1}^{i=k} v(i)^2} \qquad (1)$$

Where N is the number of samples per cycle, and $v(i)$ the recorded voltage waveform as shown in *Figure 2*. Applying (1) to *Figure 2* results in the rms voltage versus time as shown in *Figure 3*. This figure shows how the rms value drops from its pre-event value to a value of about 20% where it stays for almost 2 cycles after which it recovers to a value somewhat below its pre-event value. The full recovery takes place slowly, partly outside of the measurement window in this case. Note that the transition from pre-event to during-event voltage takes about 1 cycle in *Figure 3* where it takes place almost instantaneous in *Figure 2*. This difference is due to the 1-cycle window used in (1). Using a half-cycle window will give a faster transition but will give oscillations in the rms value when a second-harmonic component is present.

The voltage dip magnitude is the value of the remaining voltage during the deep part of the dip, typically the lowest value obtained is used. The dip duration is typically defined as the amount of time during which the rms voltage is below a certain threshold. A commonly used value for this threshold is 90%, but other values are also in use.

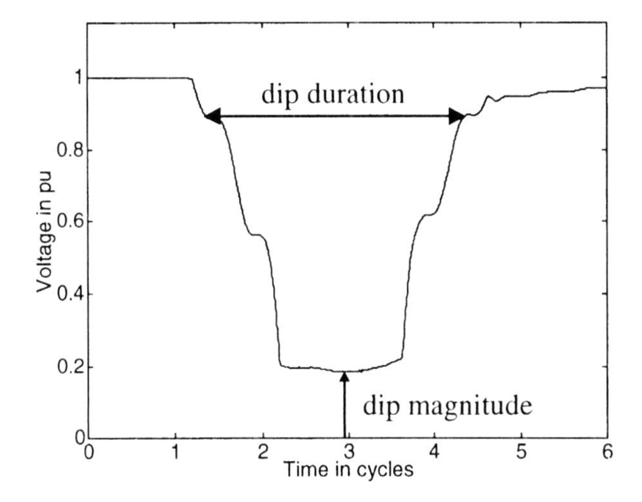

Figure 3. Example of a voltage dip: rms voltage versus time. Data obtained from [10].

The discussion about the definition of dip magnitude and duration is ongoing, but for the moment the above definitions are most-commonly used. Some problems with the definition of magnitude and duration will be given below. Even though it may not be possible to define the characteristics of individual events, it is still possible to define the supply performance in terms of dip magnitude and duration. This will be discussed further in Section 2.5.

2.2 Origin of Voltage Dips

2.2.1 Short-circuit and Earth Faults

The most important cause of voltage dips are short-circuit and earth faults. These events lead to the most severe drops in voltage and to the majority of equipment trips. An example of a voltage dip due to a fault is shown in *Figure 4*. For this and the following figures, the rms voltage as a function of time is given for the three phases. A one-cycle window has been used in all cases. The event in *Figure 4* is associated with different voltage drops in two phases whereas the rms voltage in the third phase is not affected by the event. Different rms voltages in the three phases are a common phenomenon with voltage dips due to faults. Only three-phase short-circuits lead to equal rms voltages in the three phases. Another noticeable characteristic of the event, shown in *Figure 4*, is the sharp drop and recovery of the voltage: associated with fault initiation and fault clearing, respectively. Where this figure shows a voltage dip, with a well-defined magnitude in each phase and well-defined event duration, many voltage dips due to faults are of a multi-stage character.

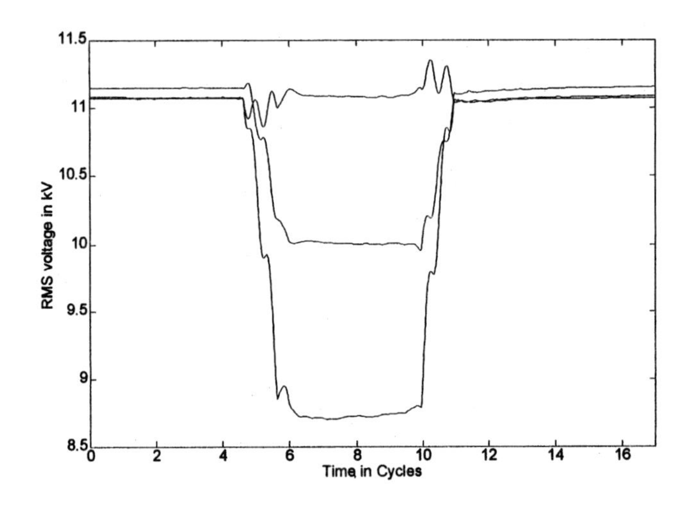

Figure 4. Voltage dip due to an unsymmetrical fault.

An example of a multi-stage voltage dip is shown in *Figure 5*.

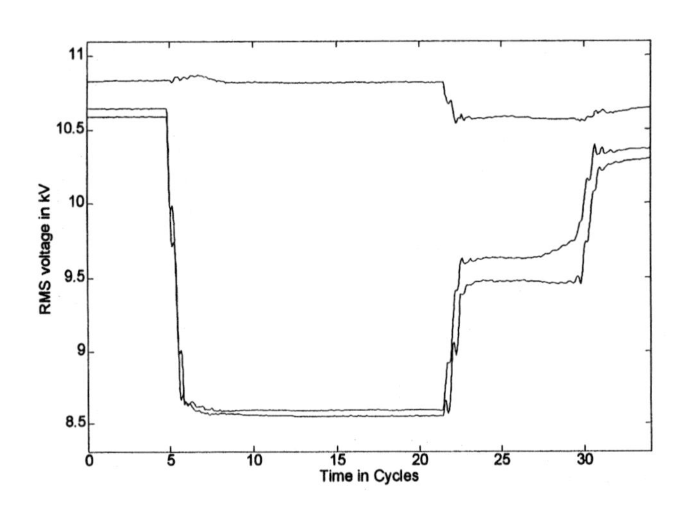

Figure 5. Multi-stage voltage dip due to a fault

It is no longer possible to characterize the event by one magnitude for each phase and by one duration. *Multi-stage voltage dips* can be due to the following phenomena:

- a change in fault type, *e.g.* from a single-phase to a two-phase fault;
- clearing of a fault by two or more breakers at different moments in time A typical example is the clearing from one side of a transmission line in zone 1 and from the other side in zone 2 of the distance relay.
- big changes in fault resistance.

2.2.2 Motor Starting and Transformer Energizing

Another common cause of voltage dips is the starting of loads associated with large inrush currents. Without counter-measures almost all loads will take a higher current with starting than during normal operation. Even a simple load such as a television will take a current of 20 to 50 A, when connected to the wall outlet. However this over-current is of very short duration; rarely longer than one ms so that the net effect on other equipment is small. Also the current may be large compared to the rated current of the television, it is still small compared to the short circuit of the system. A much more serious effect on the system voltage is due to the starting current of large induction motors. The duration of the over-current is much longer; also the over-current is a substantial part of the short circuit current of the system. An example of a voltage dip due to induction motor starting is shown in *Figure 6*.

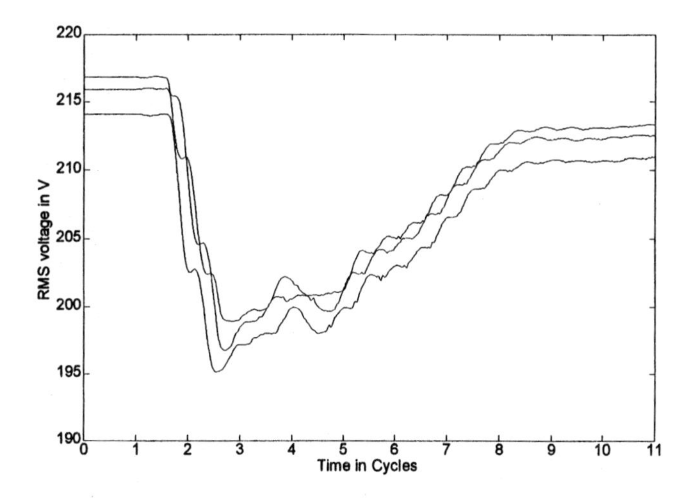

Figure 6. Voltage dip due to motor starting

Dips due to motor starting show a sharp drop in voltage but a slow recovery. The shape of the voltage recovery depends on the speed-torque relation of the mechanical load. As an induction motor is a balanced load, the over-current is the same in the three phases and thus the drop in rms voltage is the same in the three phases.

Also transformer energizing leads to a large inrush current and thus to a voltage dip, and example of which is shown in *Figure 7*.

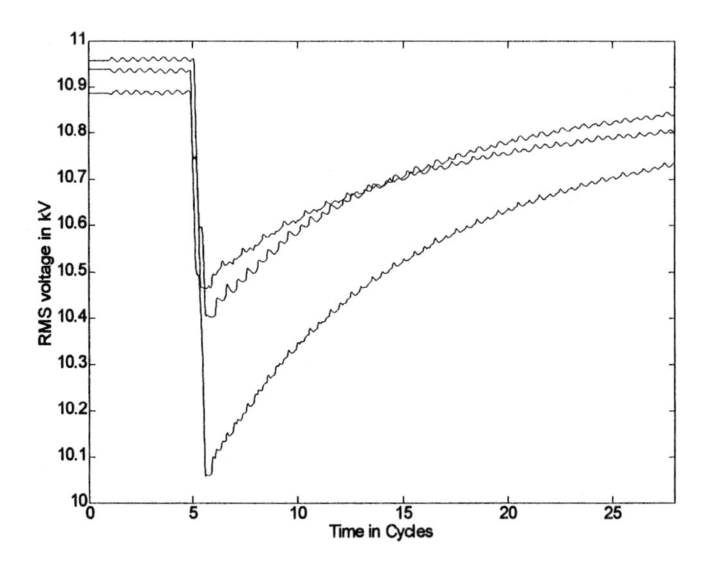

Figure 7. Voltage dip due to transformer energizing

The energizing of a transformer leads to a transient in the magnetic flux through transformer core. The flux transient contains an exponentially damped term, similar to the "dc component" that appears when switching a inductor. This "dc flux component" leads to saturation of the transformer core once every cycle, resulting in current peaks once every cycle. This can be described in time-domain as a repetition of current peaks. The result is a repetition of short voltage dips (a few ms duration). A more suitable description is a voltage dip (over-current) with a large amount of even harmonic distortion. The recovery of the voltage is related to the decay of the dc component in the magnetic flux through the transformer core. The characteristics of voltage dips due to transformer energizing are a sudden drop in voltage; a slow (exponential) recovery of the voltage; different drops in rms voltage in the three phases; the presence of even harmonics associated with transformer saturation [14].

2.2.3 Post-fault Voltage Dips

A fault is a very severe event in a power system, affecting both system and load often to a very serious extent. Upon voltage recovery, the system is no longer the same as before the fault. In severe cases, the fault may even lead to stability problems, but even for rather mild events the system recovery makes that the voltage remains below its pre-event value longer than the actual duration of the fault. The two main causes for this "post-fault voltage dip" are induction motor re-acceleration and transformer saturation. An example of a post-fault dip due to induction motor re-acceleration is shown in *Figure 8*. During the fault, the induction motors slows down, thus taking gradually more current from the supply. This increased load current leads to further drop in rms voltage. This effect is clearly visible in *Figure 8*.

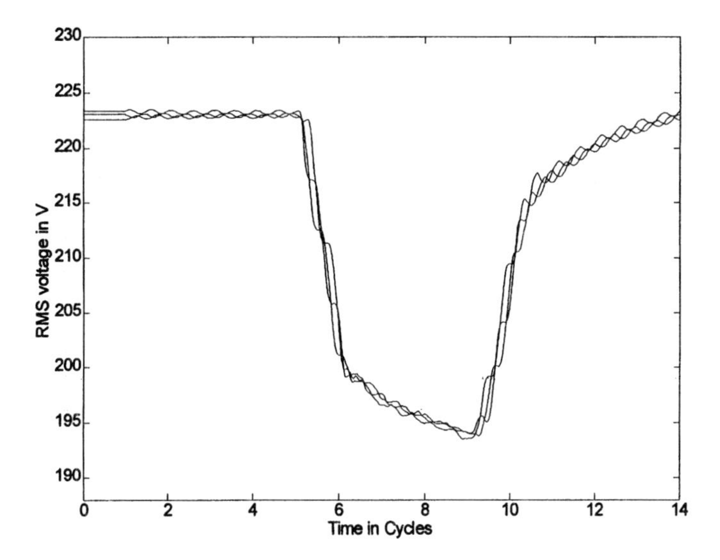

Figure 8. Voltage dip due to a three-phase fault with post-fault dip due to induction motor re-acceleration.

Upon voltage recovery the motors start to re-accelerate but as long as their speed is less than the pre-event speed, they take more current from the supply. This additional post-fault current causes the post-fault voltage dip, as clearly shown in the figure. Note that the post-fault voltage dip due to induction motor re-acceleration is balanced, *i.e.* the voltage drop is the same in the three phases. The post-fault voltage dip is most severe after a three-phase fault but it is also present after many unsymmetrical faults. The post-fault dip due to motor recovery may be more severe than a voltage dip due to motor starting. With motor starting only one motor takes a high current. A voltage dip effects all motors, so that they all take a high current when the voltage recovers. This effect is a limiting factor in the amount of motor load that can be fed from one bus in the system.

An additional effect of the voltage recovery after a fault is that transformers will saturate. Especially non-loaded and lightly loaded transformers are prone to this effect, but even heavily loaded transformers have shown this phenomenon. The saturation of transformers upon voltage recovery will lead to a post-fault voltage dip of the same character as the voltage dip due to transformer energizing. An example of such an event is shown in *Figure 9*. The post-fault voltage dip due to transformer energizing is characterized by unbalance and presence of even harmonics [13]. A post-fault dip due to induction-motor recovery occurs mainly in low-voltage and distribution networks. A post-fault dip due to transformer saturation occurs at all voltage levels.

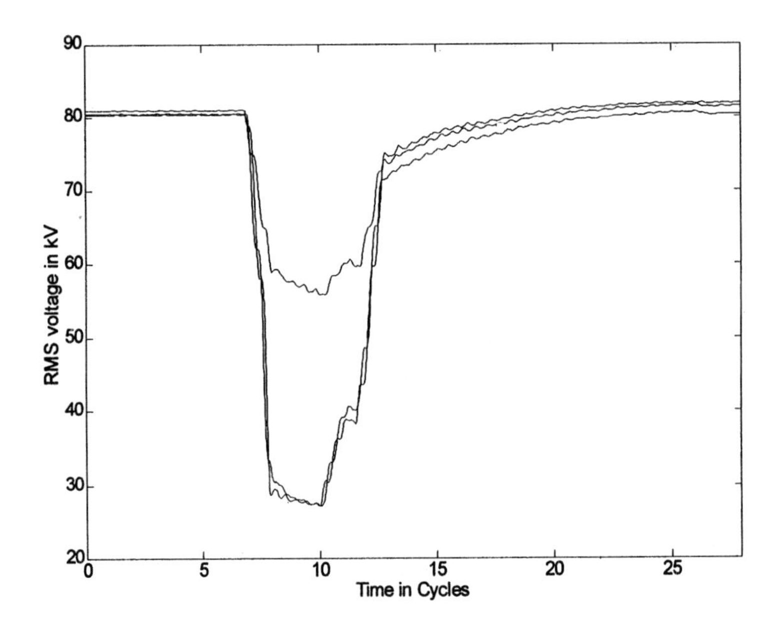

Figure 9. Voltage dip followed by a post-fault dip due to transformer saturation

2.3 Magnitude, Duration and Fault Location

Faults at different locations lead to different sag characteristics. Both magnitude and duration are affected by the fault location. For the load in

Figure 10, a fault at position 1 will cause a drop in voltage. Fault-clearing times in low-voltage networks are typically short: current-limiting fuses and miniature circuit breakers are typical forms of fault clearing. Further, the dip frequency due to low-voltage faults will be low for a number of reasons. Low-voltage networks are often underground with the associated small fault frequencies. The number of kilometers of low-voltage network, to which a specific load is exposed, is small since low-voltage networks are restricted to small areas. Faults in remote low-voltage networks (position 3 in *Figure 10*) do not lead to any significant dip due to the large impedance of the distribution transformer.

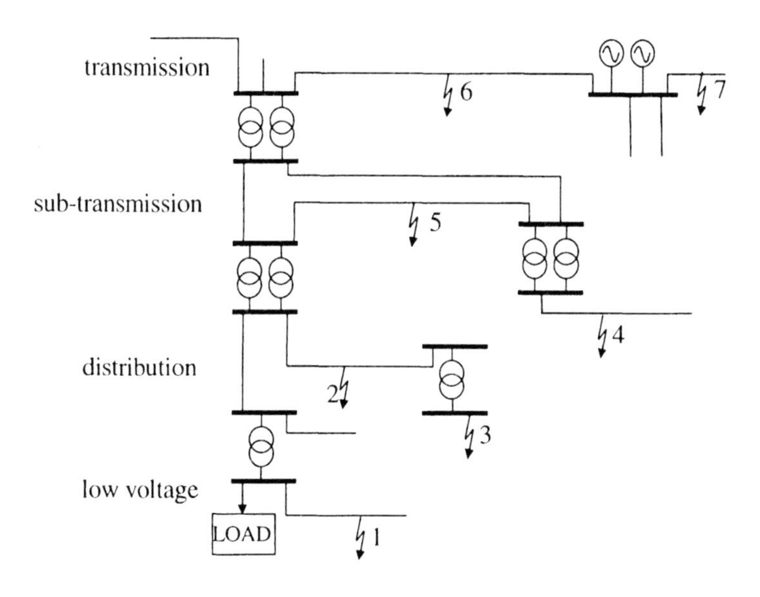

Figure 10. Supply network with different fault locations

Of much more concern are dips due to distribution system faults (locations 2 and 4). Not only are fault-clearing times much longer (up to 5 seconds in some systems) but also is the load exposed to many kilometers of feeder and are these feeders often overhead. Faults in a remote distribution network (location 4) do lead to a smaller drop in voltage, but the long duration may still pose a concern to sensitive equipment. Faults at sub-transmission and transmission level (locations 5, 6 and 7) lead to dips of any magnitudes depending on the location, with a relatively short duration; typically around 100 ms. The voltage drop becomes less when the fault is located further away from the load, but a fault hundreds of kilometers away may still lead to a serious drop in voltage (*e.g.* down to 70% of nominal). A fault behind a large power station (location 7) no longer poses a threat to the load. The power station operates as a voltage source keeping up the voltage at the station bus and thus for all customers fed from this bus.

An overview of the different groupings in the magnitude-duration plane is shown in *Figure 11*. Current-limiting fuses cause very short dips; transmission system faults cause short dips while distribution system faults cause long dips. Motor and transformer switching leads to long events. For completeness also interruptions are included in the figure, as long deep events.

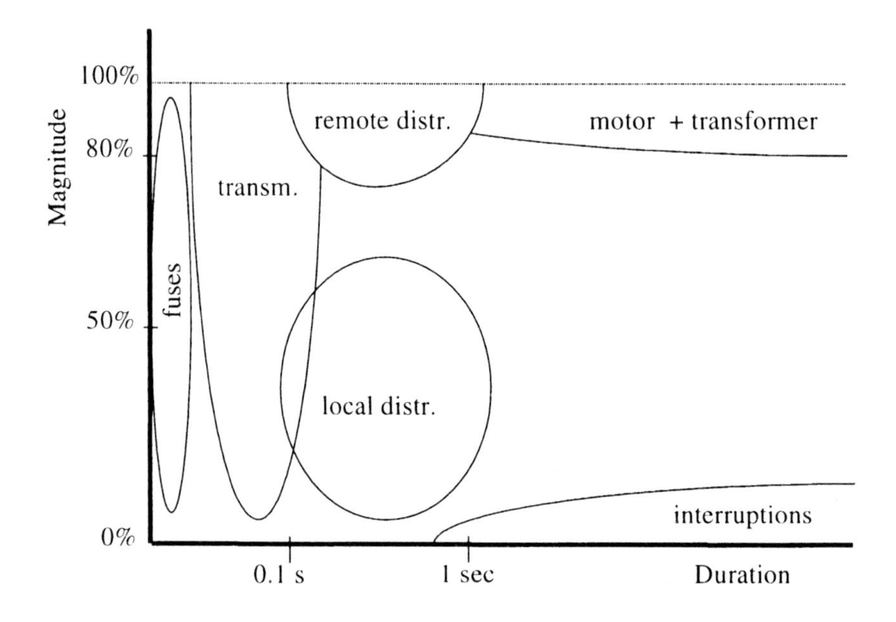

Figure 11. Dips of different origin in the magnitude-duration plane

2.4 Three-Phase Unbalanced Dips

In section 2.1 the dip magnitude was introduced as a characteristic of a single-phase event. To come to performance indices of the supply, it is important that dip magnitude is defined in an appropriate way. The definition should be such that it can be applied to equipment behavior. The characteristics of an event should give a good indication of the equipment behavior that can be expected due to the event. For three-phase unbalanced dips, as shown for example in *Figure 4*, *Figure 5* and *Figure 9*, it is no longer trivial to define one single dip magnitude.

2.4.1 Basic Classification

To characterize the dip as experienced by three-phase equipment, it is needed to use the method of symmetrical components. It can be shown that three-phase unbalanced voltage dips come in three types: designated as type A, type C and type D. Type A is a balanced dip due to a three-phase fault:

$$\begin{aligned}\overline{V}_a &= \overline{V}\\ \overline{V}_b &= -\tfrac{1}{2}\overline{V} - \tfrac{1}{2}j\overline{V}\sqrt{3}\\ \overline{V}_c &= -\tfrac{1}{2}\overline{V} + \tfrac{1}{2}j\overline{V}\sqrt{3}\end{aligned} \tag{2}$$

with $\overline{V}$ the "characteristic voltage" of the event. For a type C dip the expressions are:

$$\begin{aligned}\overline{V}_a &= \overline{F}\\ \overline{V}_b &= -\tfrac{1}{2}\overline{F} - \tfrac{1}{2}j\overline{V}\sqrt{3}\\ \overline{V}_c &= -\tfrac{1}{2}\overline{F} + \tfrac{1}{2}j\overline{V}\sqrt{3}\end{aligned} \tag{3}$$

and for a type D dip:

$$\begin{aligned}\overline{V}_a &= \overline{V}\\ \overline{V}_b &= -\tfrac{1}{2}\overline{V} - \tfrac{1}{2}\overline{F}j\sqrt{3}\\ \overline{V}_c &= -\tfrac{1}{2}\overline{V} + \tfrac{1}{2}\overline{F}j\sqrt{3}\end{aligned} \tag{4}$$

with $\overline{F}$ the so-called "PN-factor" of the event. The characteristic magnitude plays the same role as the magnitude (remaining voltage) for single-phase events. The PN-factor can be seen as a correction factor indicating how much the event deviates from an "ideal three-phase unbalanced voltage dip". Measurements have shown that the PN-factor is close to unity for dips due to single-phase and phase-to-phase faults. A small drop in PN-factor is observed with measurements. The drop somewhat increases during the course of the event. Simulations have shown that this can be explained from the reduction in speed of induction motor load. For two-phase-to-ground faults the PN-factor is significantly smaller than unity. This characterization is discussed in detail in [16 and 17].

The effect of characteristic voltage and PN-factor on the three voltage phasors is shown in *Figure 12*. For a type C dip, the characteristic voltage relates to the remaining voltage between the two most-severely-affected phases; the PN-factor is the remaining voltage in the third phase. For a type

D event, the characteristic voltage is the remaining voltage in the most-severely-affected phase, where the PN-factor relates to the remaining voltage between the two other phases.

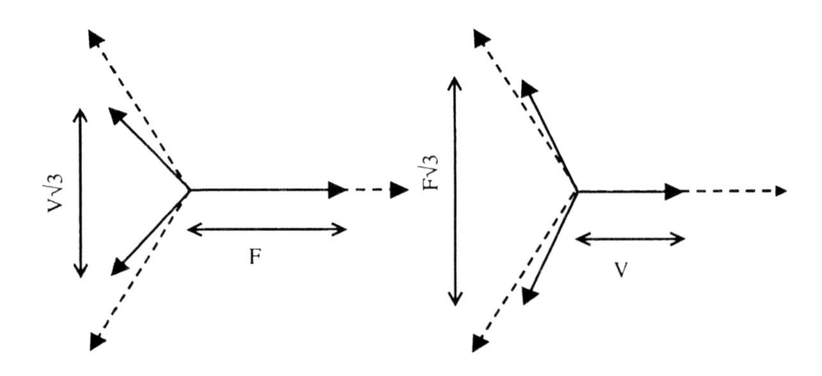

Figure 12. The two basic types of three-phase unbalanced dips: type C (left) and type D (right). The pre-event voltages are indicated through dashed lines.

2.4.2 Symmetrical Phase

Expressions (3) and (4) are valid for a fault in phase a or between phases b and c, i.e. with phase a as the symmetrical phase. Obviously faults may occur between any phase and ground, and between any pair of two phases. In other words: there are three different symmetrical phases. Including these three possible symmetrical phases results in six (sub)-types of three-phase unbalanced dips: C_a, C_b, C_c and D_a, D_b, D_c. Type D_b is a type D dip with b as symmetrical phase: thus a voltage drop in phase b with minor drops in phases a and c. Type C_b is a dip op type C with b as symmetrical phase: a drop of the voltage difference between phases b and c. Expression (3) describes a dip of type C_a; expression (4) a dip of type D_a. The expressions for the other four types can be obtained through rotating the voltages in the complex phase over 120°. The six types are shown in phasor-diagram form in *Figure 13*.

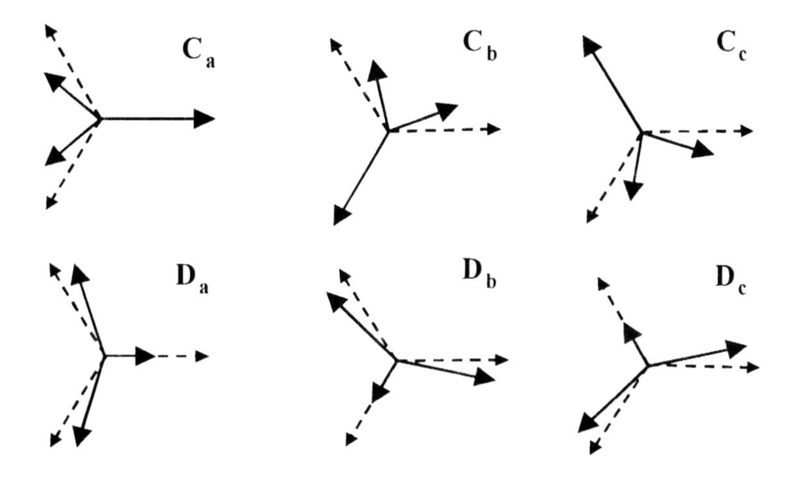

Figure 13. Six types of three-phase unbalanced dips, in phasor-diagram form, when considering the symmetrical phase

2.4.3 Example

An example of an unbalanced dip is shown in *Figure 14*. The plot shows rms voltage versus time for the three phases. The rms voltage is calculated according to expression (1) by using a one-cycle window. The event shows a severe drop in two phases and a small drop in the third phase. Note also that the voltage in the third phase decays during the event and doesn't show any change upon fault clearing. By using this phase only, it would thus be difficult to detect beginning and end of the dip.

From the voltage wave-shapes (not shown here) the complex phase voltages as a function of time are calculated. These are used to calculate positive and negative-sequence voltages according, which are in turn used to determine dip type and characteristics. The method used to calculate voltage dip type and characteristics from measured complex phase voltages is described in detail in [16 and 17]. An alternative method, using only rms voltages, is described in [1]. Recent studies, not yet published, have shown that this simple method gives an incorrect result in some cases but that it could be used in case statistical results are sought after so that an error in a small number of cases is not a problem.

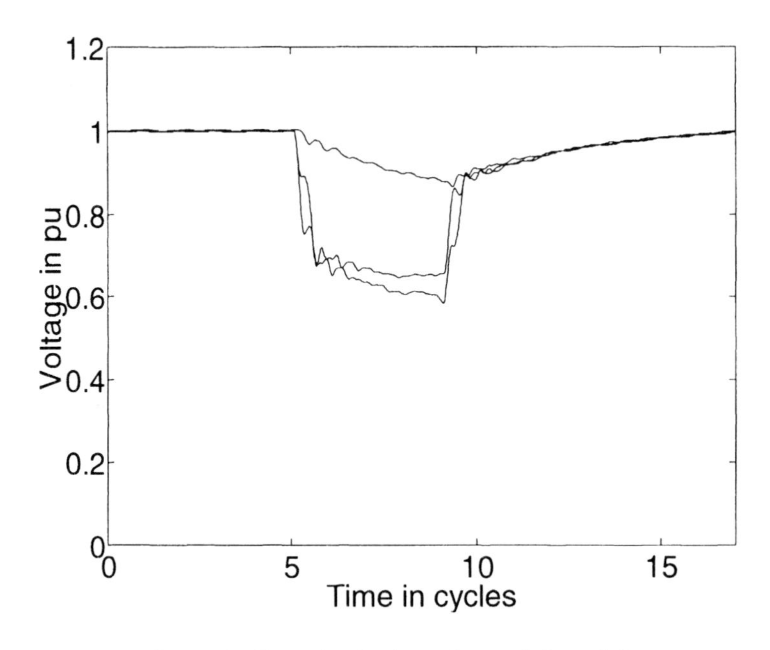

Figure 14. Example of a three-phase unbalanced dip

The resulting characteristic magnitude (the absolute value of the characteristic voltage) and the absolute value of the PN factor are shown in *Figure 15*. The characteristic magnitude shows the same kind of behavior as the rms voltage (the "magnitude versus time") for single-phase events. (See *e.g.* Figure 3). The PN factor shows a slow but steady decrease during the fault and a slow but steady increase after the fault. Neither at fault initiation, nor at fault clearing does the PN-factor show any sudden change. This behavior has been shown to be typical for many recorded voltage dips. The characteristic magnitude is in all cases "a real dip in voltage". The PN factor is sometimes constant, and its decay shows different shapes. But the PN factor never increases during the fault, and the PN-factor never shows a jump at fault initiation or at fault clearing. The behavior of the PN-factor can be explained from the slowing down and speeding up of induction motor load during and after the fault. Not shown in Figure 15 is the behavior of the phase-angle of the PN-factor. Also the phase-angle shows a continuous behavior without any sudden jumps.

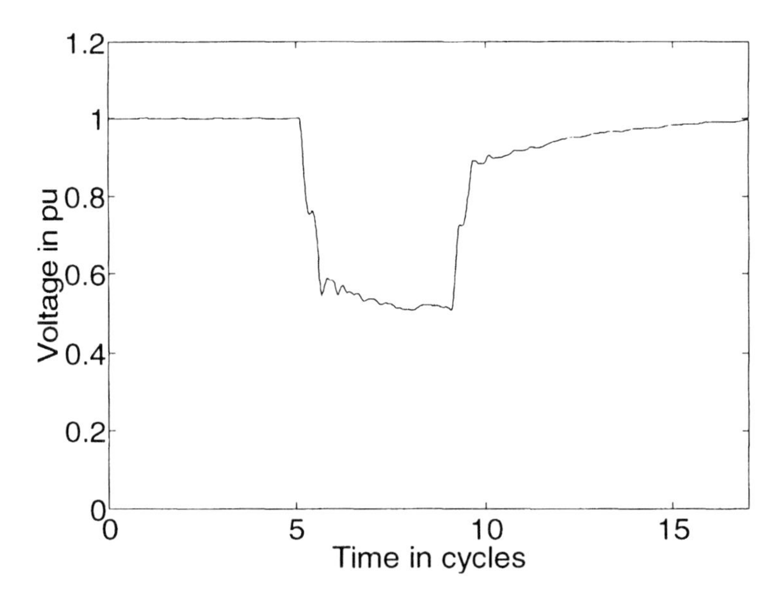

Figure 15. Characteristic magnitude (solid line) and absolute value of the PN factor (dashed line) for the dip shown in *Figure 14*

2.5 Voltage Dip Performance Indicators

All the discussion in the previous parts of the chapter has been about individual events. What is of much more use, to both customer and distribution or transmission company, are statistics on the number of events with certain characteristics. Statistical indices are needed in, among others, the following situations:

- When comparing the dip performance of different distribution and transmission companies. As mentioned in the previous chapter, deregulation and privatization require a greater transparency on the supply performance. Next to the publication of interruption statistics, it is also desirable that distribution and transmission companies publish statistics on voltage dips. The role of the independent system operator or regulatory body is to define suitable performance indicators and possibly even performance requirements.

- Certain countries and cities attract industry through a highly reliable infrastructure, including electricity. To back-up this, distribution and transmission companies may want to publish voltage dip performance indicators.

- In power quality contracts between distribution or transmission companies and customers. Any guarantee on power quality requires a well-defined voltage dip performance indicator or even a set of indicators.

- When assessing the need for the mitigation measures against voltage dips. Individual customers may want to know the quality of the supply to decide if it is worth to invest in mitigation measures.

2.5.1 Voltage-Dip Co-ordination Chart

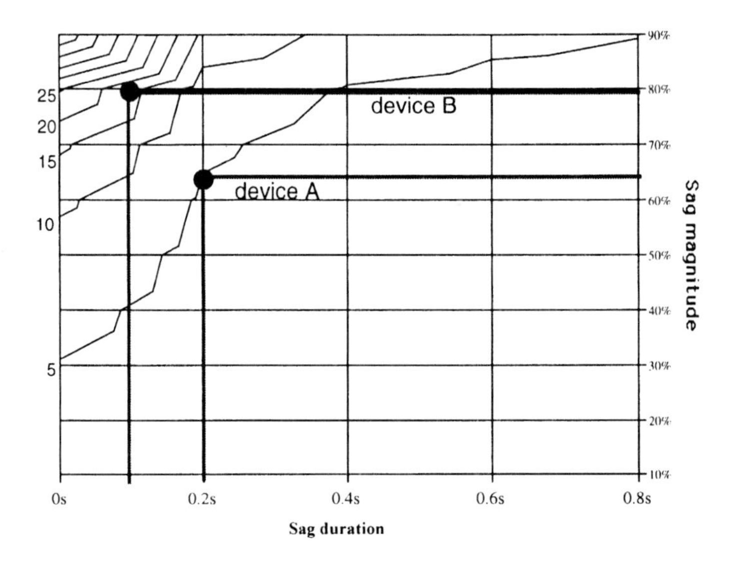

Figure 16. Example of a voltage dip co-ordination chart: number of events per year

There are a number of ways to characterize the supply performance for voltage dips. Most proposed methods are based on magnitude and duration of the event. A very useful method is through the voltage-dip co-ordination chart [7], an example of which is shown in *Figure 16*. The underlying idea is the creation of a two-dimensional function: the dip frequency as a function of magnitude and duration. Each function value is the (predicted or measured) number of events whose magnitude and duration are more severe than the given value. This function is represented through a contour chart. For the supply of *Figure 16*, there are 5 events per year with a duration exceeding 200 ms and a magnitude below 65% of nominal. The voltage-sag co-ordination chart is directly related to equipment behavior. Drawing the voltage-tolerance curve of the equipment will almost immediately give the expected number of equipment trips. In this example: device A will trip about 5 times per year, and device B about 18 times. Simple rules can be derived for more complicated (non-rectangular) voltage-tolerance curves, but most equipment actually possesses a rectangular curve.

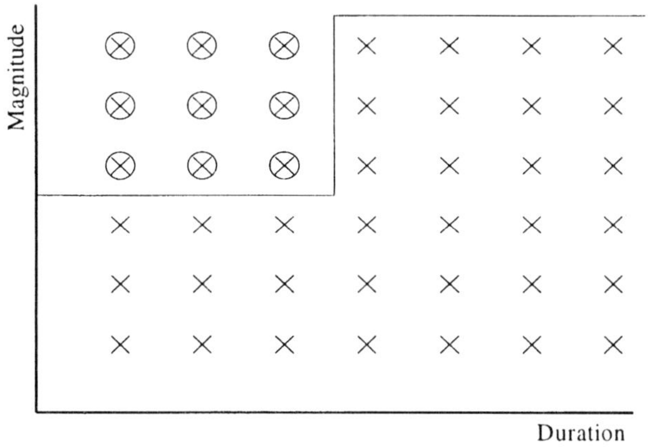

Figure 17. Update of voltage-dip co-ordination chart for rectangular voltage dip. The cross with circle indicates those values for which the dip frequency is increased.

2.5.2 Non-rectangular Events

In Section 2.2 a number of examples were shown of voltage dips in which the magnitude and duration could not be easily uniquely defined. For these non-rectangular dips the voltage-dip co-ordination chart can still be used, even though it is not possible to characterize individual events. This is done by re-interpretation of the underlying two-dimensional function: the dip frequency as a function of magnitude and duration. Originally, the function value was defined as the number of events exceeding the given severity.

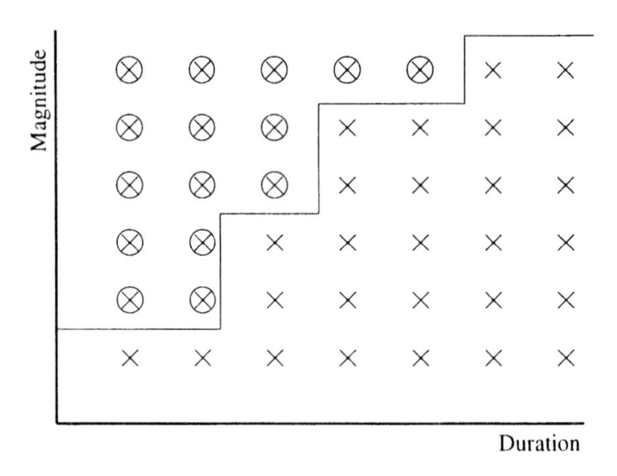

Figure 18. Update of voltage-dip co-ordination chart for non-rectangular voltage dip

Alternatively, the function value can be defined as the number of times per year that the rms voltage is less than the given magnitude for longer than the given duration. For the rectangular dip shown as rms voltage versus time in *Figure 17*, the function value is increased by one for the indicated values of magnitude and duration (cross with circle). The function value for the other elements is not affected. This way of increasing the function value can be applied to any shape of the rms voltage as a function of time, also to a non-rectangular shape as in *Figure 18*. For a non-monotonous increasing voltage dip shape, as shown in *Figure 19*, the magnitude-duration values are shifted towards the left of the plane before the function value is increased. This method is described *e.g.* in reference [11]. The method proposed in reference [4] is based on the same line of reasoning: what mattes is not the actual shape of individual events, but the number of times per year that the voltage drops below a given threshold for longer than a given duration. This is sometimes explains as saying that the duration of the event depends on the threshold used. A lower threshold value will result in a shorter event.

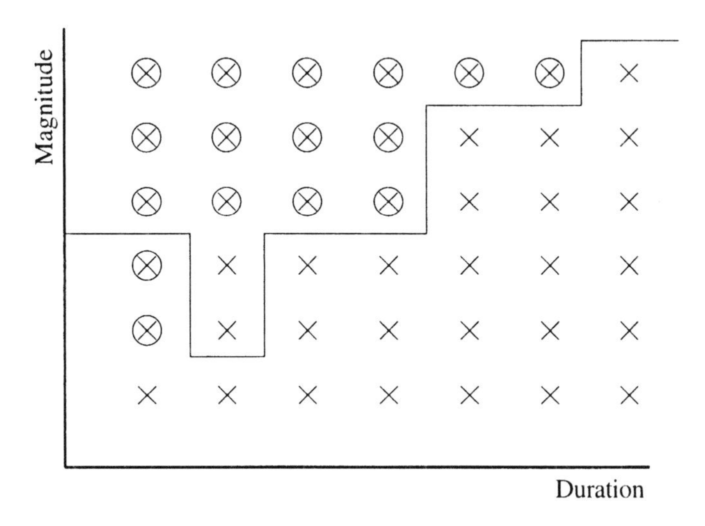

Figure 19. Update for voltage-dip co-ordination chart for multi-stage voltage dip

2.5.3 Three-phase Unbalance

For a three-phase supply it is appropriate to include all three-phase voltages in any performance indicator. This holds especially for voltage dips as the majority of them are unbalanced. Some of the existing proposals use the lowest of the three phase voltages as the magnitude of the event, before coming to an indicator like the voltage-dip co-ordination chart.

The classification of three-phase unbalanced dips as discussed in Section 2.4, is a more appropriate choice. The characteristic magnitude together with the duration, or as function of time for non-rectangular events, can be used to obtain a voltage-dip co-ordination chart for each of the three types. An example is shown in *Figure 20*, with voltage-tolerance curves indicated. Note that the voltage-tolerance curves will typically be different for the three dip types. The total number of equipment trips, for this example, will simply be the sum of the values obtained from the three dip types:

$$N_{trip} = N_A + N_C + N_D \tag{5}$$

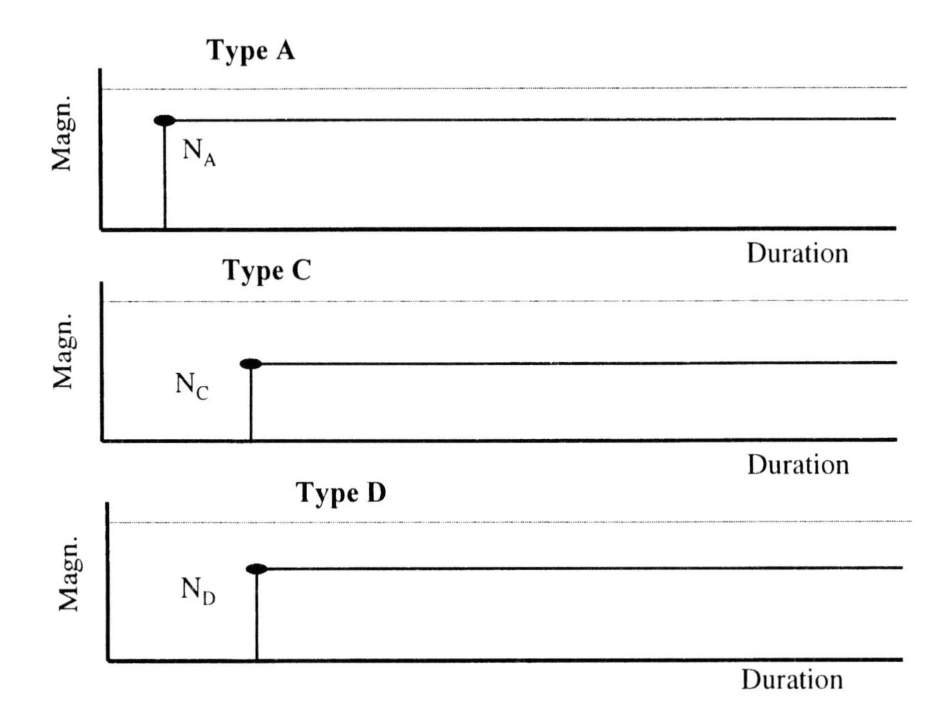

Figure 20. Voltage dip co-ordination in three phases

2.5.4 Remaining Problems

A number of problems remain to be solved in the development of suitable voltage-dip performance characteristics. For three-phase supplies, the PN factor is not part of the proposed performance characterization method. A dip with a small PN factor value may be more severe than a dip with the same characteristic magnitude and with a unity PN factor. In reference [2] it was shown that this is exactly the case for ac adjustable-speed drives.

A few other dip characteristics have been introduced by various authors. Next to magnitude and duration, phase-angle jump and point-on-wave of dip initiation have been shown to affect equipment behavior. Including point-on-wave of dip initiation is probably relatively easy as this variable is likely to be independent of magnitude and duration. Including PN-factor and phase-angle jump may be more difficult. It will lead to a multi-dimensional dip frequency function that no longer allows any easy interpretation. Further work, theoretical modeling and analysis of monitoring results, is needed here to obtain relations between magnitude, duration, phase-angle jump and PN factor, that bring back the multi-dimensional problem to one or more two-dimensional problems.

The voltage-dip co-ordination chart and the other performance indicators proposed, give a (large) number of values for the supply performance. In some cases one is interested in obtaining one single value, or in the worst case a small number of values. Compare this with system reliability, where the supply performance may be characterized through the interruption frequency and the unavailability. It is obvious that this will lead to the loss of relevant information, *e.g.* the distribution function of the interruption duration, but it enables a fast assessment of the reliability of the supply. No such indices have been proposed yet for the voltage dip performance of the supply.

2.5.5 Data Collection

With any voltage dip performance indicator, it is needed to collect data on voltage dip frequency. The two basic methods for this are monitoring and stochastic prediction. With voltage dip monitoring, measurement equipment is installed at a number of locations in the system. For each occurring event the characteristics (magnitude, duration, etc.) are determined and stored. After a certain period, statistics are obtained resulting in the above-mentioned voltage-dip co-ordination chart. Several large surveys have been held to obtain the "average power quality" for a certain region or country. The large survey held by EPRI in the US is the most well-known [8], but also elsewhere large surveys have been conducted, *e.g.* [9]. Such a survey typically lasts a number of years and the results do give a kind of average power quality as well as a benchmark for distribution and transmission companies to compare their performance against. These surveys have also taught us a lot about power quality phenomena and the resulting databases will keep researchers busy for years to come. However the usefulness as a performance indicator is limited. At the most, one can compare the power quality in different countries with different design and operational practices.

To obtain useful performance values, the measurements should be collected over a shorter period and limited to one distribution or transmission company. By annually publishing the results, a certain comparison of the voltage dip performance can be made. Obtaining the data becomes somewhat tricky, in this case. There are a number of ways to collect the data, each with obvious advantages and disadvantages:

- A very large number of monitors can be installed, *e.g.* with every customer or with each distribution transformer. This will give the most accurate performance data, but the investment in monitoring equipment is probably too big for any company to even consider.

Future electronic metering may include a limited form of power quality monitoring. The same holds for future generations of numerical protection relays. But collecting the data still does not solve the communication and data processing problem.

- An alternative is to install a limited number of monitors (*e.g.* 100) randomly spread over the service territory of the distribution or transmission company. The "law of the large numbers" in statistics will (hopefully) make that the average results are the same as when a very large number of monitors would be installed. The disadvantage of this method is that one never knows if the results are the same, and that the investment in communication and data collection may still be very large. An additional disadvantage is that it is no longer possible to obtain performance indicators for individual customers. This may make it difficult to give performance guarantees to customers.

- One may go even a step further and only install at a small number of locations (*e.g.* 10) spread through the network. The choice of the monitor locations should be such that a fair comparison with other companies is possible. The difficulties in choosing such locations are the obvious disadvantages of this method.

- By using some knowledge about the system, it may be possible to obtain information for all locations from a small number of locations. Knowing voltages and currents from a limited number of locations, it should be possible to estimate the voltages throughout the system. Information from protection relays may be used as additional input information. Methods for this *power-quality state estimation* need to be further developed, and verified. If the estimation is accurate enough, full covering of the system can be obtained without the need to install monitors at many locations.

Next to all the above monitoring-based methods, *stochastic prediction methods* can be used to estimate system performance. Stochastic voltage dip prediction operates in the same way as power system reliability analysis: from past-performance statistics, the expected future behaviour of the system is calculated. Techniques for this are discussed, among others, [3 below], [6], and [11]. The advantage of stochastic prediction is that it does not require expensive and time-consuming monitoring programs, that it is very easy to distinguish between dips with different causes and originating at different locations, and that the result is not affected by year-to-year variations. The latter is also the main disadvantage of the method: it is not possible to compare the year-to-year performance. An alternative that does not have this disadvantage is to use the fault statistics of one specific year and calculate the resulting voltage dip statistics from these.

2.6 The Responsibility Question

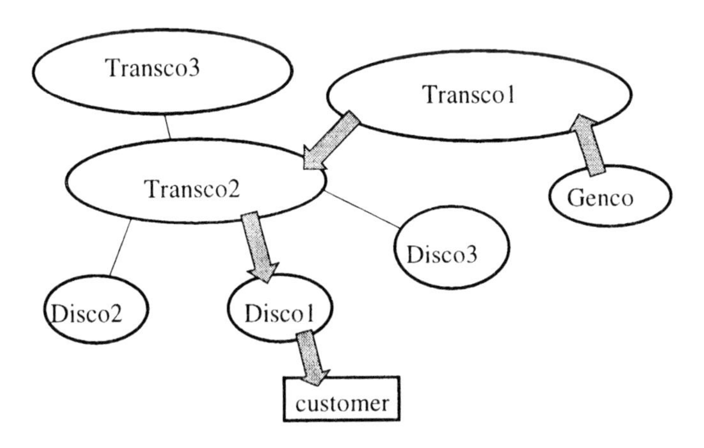

Figure 21. Flow of electrical energy through the power system from a generator through transmission and distribution networks to a specific customer

Voltage dip performance statistics and data-collection methods have been discussed above, but the critical question still hasn't been touched: *who is responsible for the voltage dip performance?* The deregulation of the electricity industry has not simplified this question. Consider as an example the situation shown in *Figure 21*: a certain customer buys electrical energy from a certain generation company (Genco). To transport and deliver the energy, two transmission companies (Transco1 and Transco2) and one distribution company (Distco1) need to be involved. The various types of contracts between these 5 partners have been discussed in previous chapters, here the discussion will be on voltage dips. *Figure 22* shows the origin of voltage dips experienced by the customer. The Genco does not contribute to the voltage dip frequency, but all the transmission and distribution companies do. Transco1, Transco2 and Distco1 are partners in this transaction, so that their responsibility could be somewhat contractually defined. The problem gets more complicated with Transco3, Distco2 and Distco3. A short circuit in the network of Distco2, may cause a long dip in the network of Transco2, which propagates down to the customer connection. To contractually define the responsibility for such events is all but impossible.

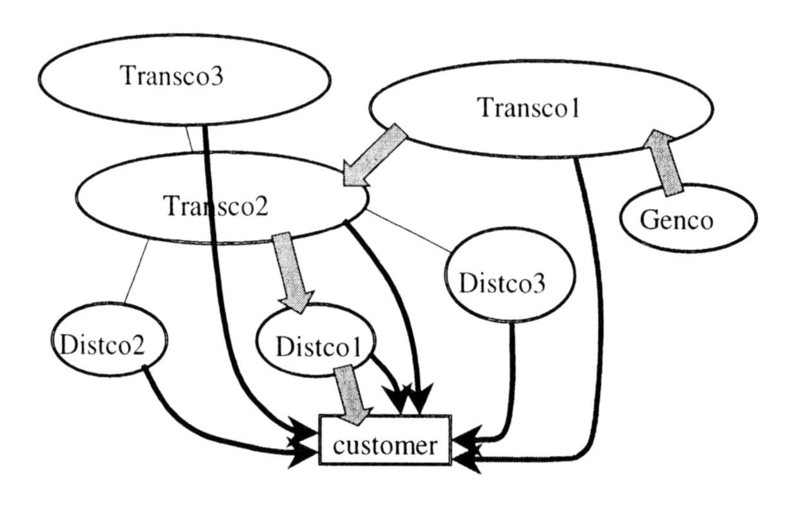

Figure 22. Different origins of dips experienced by a specific customer

The legal discussion about responsibility can be brought back to a technical discussion about the origin of voltage dips. Whether a distribution or transmission company is actually responsible for dips originating within their system, is a question that falls outside of the realm of this chapter. But independent of the responsibility question, performance indices may be obtained considering the origin of the event.

The performance indicator discussed in Section 2.5, considered any voltage dip equally. Strictly speaking this only leads to a performance indicator of the power system as a whole, but it does not necessarily give any information about the performance of the transmission or distribution company. Voltage dips measured at a certain location may originate in a network operated by another company. One way of solving this is to include the direction of a voltage dip with its characteristics. By measuring both voltage and current it may be possible to detect in which direction a fault occurs. Knowing the direction of the dip at the interfaces with neighboring systems, external events can be excluded from the statistics. Using communication between the monitors would enable a real-time decision, but that is rarely needed. Adding a time stamp to the voltage dip will enable post-processing of all the events collected over a certain period.

Voltage dips due to induction motor starting and due to transformer energizing can be removed from the statistics and treated differently by using their voltage characteristics as discussed in Section 2.2.

A final problem that may need to be solved is the treatment of post-fault dips due to motor recovery and transformer energizing. The simplest solution is to consider them as part of the main dip, so that the origin lays with the same company as the fault. Alternatively, one may argue that the responsibility lays with the company operating the transformer and with the customers owning the motor load. Especially in the latter case, it will be impossible to allocate one specific responsible party, but one may still decide that the responsibility for the post-fault event does not lay with the company responsible for the fault. The same may be concluded for the additional drop in voltage during a fault, due to the slowing down of induction motor load. The PN-factor, as introduced in Section 2.4.1, may be used as an indicator for this drop.

From the above discussion it will be clear that it is in principal possible to set up voltage-dip performance indicators for transmission and distribution companies, but that such requires more than simply monitoring. The contractual and legal problems however, may proof more difficult.

2.7 Voltage Dip Mitigation

2.7.1 From Fault to Trip

To understand the various ways of mitigation, the mechanism leading to an equipment trip needs to be understood. The equipment trip is what makes the event a problem; if there are no equipment trips, there is no voltage dip problem. The underlying event of the equipment trip is a short-circuit fault. At the fault position the voltage drops to zero, or to a very low value. This zero voltage is changed into an event of a certain magnitude and duration at the interface between the equipment and the power system. The short-circuit fault will always cause a voltage dip for some customers. If the fault takes place in a radial part of the system, the protection intervention clearing the fault will also lead to an interruption. If there is sufficient redundancy present, the short circuit will only lead to a voltage dip. If the resulting event exceeds a certain severity, it will cause an equipment trip. Based on this reasoning it is possible to distinguish between the following mitigation methods:

- Reducing the number of short-circuit faults.
- Reducing the fault-clearing time.

- Changing the system such that short-circuit faults result in less severe events at the equipment terminals or at the customer interface.
- Connecting mitigation equipment between the sensitive equipment and the supply.
- Improving the immunity of the equipment.

2.7.2 Reducing the Number of Faults

Reducing the number of short-circuit faults in a system not only reduces the dip frequency but also the frequency of long interruptions. This is thus a very effective way of improving the quality of supply and many customers suggest this as the obvious solution when a voltage dip or interruption problem occurs. The solution is unfortunately most of the time not that obvious. A short circuit fault not only leads to a voltage dip or interruption at the customer interface but may also cause damage to utility equipment and plant. Therefore most utilities will already have reduced the fault frequency as far as economically feasible. In individual cases there could still be room for improvement, *e.g.* when the majority of trips is due to faults on one or two distribution lines. Some examples of fault mitigation are:

- Replace overhead lines by underground cables.
- Use special wires for overhead lines.
- Implement a strict policy of tree trimming.
- Install additional shielding wires.
- Increase maintenance and inspection frequencies.

One has to keep in mind however that these measures can be very expensive, especially for transmission systems, and that its costs have to be weighted against the consequences of the equipment trips.

2.7.3 Reducing the Fault-clearing Time

Reducing the fault-clearing time does not reduce the number of events but can significantly limit the dip duration. The ultimate reduction of fault-clearing time is achieved by using current-limiting fuses, able to clear a fault within one half-cycle. The recently introduced static circuit breaker has the same characteristics: again fault-clearing time within one half-cycle. Additionally several types of fault-current limiters have been proposed which do not actually clear the fault, but significantly reduce the fault current magnitude and thus the voltage drop within one or two cycles. One important restriction of all these devices is that they can only be used for low and medium-voltage systems.

But the fault-clearing time is not only the time needed to open the breaker, also the time needed for the protection to make a decision. To achieve a serious reduction in fault-clearing time, it is needed to reduce any grading margins, thereby possible allowing for a certain loss of selectivity. Next to this, faster protection algorithms may be developed, making use of communication between relays. Reduction in fault-clearing time is especially needed in distribution systems.

2.7.4 Changing the Power System

By implementing changes in the supply system, the severity of the event can be reduced. Here again the costs may become very high, especially for transmission and sub-transmission voltage levels. In industrial systems, such improvements more often outweigh the costs, especially when included already in the design stage. Some examples of mitigation methods especially directed towards voltage dips are:

- Install a generator near the sensitive load. The generator will keep up the voltage during a remote dip. The reduction in voltage drop is equal to the percentage contribution of the generator station to the fault current. In case a combined-heat-and-power station is planned, it is worth considering the position of its electrical connection to the supply.
- Split busses or substations in the supply path to limit the number of feeders in the exposed area.
- Install current-limiting coils at strategic places in the system to increase the "electrical distance" to the fault. The drawback of this method is that this may make the event worse for other customers.
- Feed the bus with the sensitive equipment from two or more substations. A voltage dip in one substation will be mitigated by the in-feed from the other substations. The more independent the substations are the more the mitigation effect. The best mitigation effect is by feeding from two different transmission substations. Introducing the second in-feed increases the number of dips but reduces their severity.

2.7.5 Installing Mitigation Equipment

The most commonly applied method of mitigation is the installation of additional equipment at the system-equipment interface. Also recent developments point towards a continued interest in this way of mitigation. The popularity of mitigation equipment is explained by its being the only place where the customer has control over the situation. Both changes in the supply as well as improvement of the equipment are often completely outside of the control of the end-user. Some examples of mitigation equipment are:

- *Uninterruptible Power Supply (UPS).* This is the most commonly used device to protect low-power equipment (computers, etc.) against voltage dips and interruptions. During the dip or interruption, the power supply is taken over by an internal battery. The battery can supply the load for, typically, between 15 and 30 minutes.

- *Static transfer switch.* A static transfer switch switches the load from the supply with the dip to another supply within a few milliseconds. This limits the duration of a dip to less than one half-cycle assuming that a suitable alternate supply is available.

- *Dynamic Voltage Restorer (DVR).* This device uses modern power electronic components to insert a series voltage source between the supply and the load. The voltage source compensates for the voltage drop due to the dip. Some devices use internal energy storage to make up for the drop in active power supplied by the system. They can only mitigate dips up to a maximum duration. Other devices take the same amount of active power from the supply by increasing the current. These can only mitigate dips down to a minimum magnitude. The same holds for devices boosting the voltage through a transformer with static tap changer.

- *Motor-generator sets.* Motor-generator-sets are the classical solution for dip and interruption mitigation with large equipment. They are obviously not suitable for an office environment but the noise and the maintenance requirements are often no objection in an industrial environment. Some manufacturers combine the motor-generator-set with a backup generator; others combine it with power-electronic converters to obtain a longer ride-through time.

2.7.6 Improving Equipment Voltage Tolerance

Improvement of equipment voltage tolerance is probably the most effective solution against equipment trips due to voltage dips. But as a short-time solution it is often not suitable. In many cases, a customer only finds out about equipment performance after it has been installed. Even most adjustable-speed drives have become off-the-shelf equipment where the customer has no influence on the specifications. Only large industrial equipment is custom-made for a certain application, which enables the incorporation of voltage-tolerance requirements in the specification.

Apart from improving large equipment (drives, process-control computers) a thorough inspection of the immunity of all contactors, relays, sensors, etc. can significantly improve the voltage tolerance of the process.

2.7.7 Different Events and Mitigation methods

Figure 11 showed the magnitude and duration of voltage dips and interruptions resulting from various system events. For different events different mitigation strategies apply.

Dips due to short-circuit faults in the transmission and sub-transmission system are characterized by a short duration, typically up to 100 ms. These dips are very hard to mitigate at the source and also improvements in the system are seldom feasible. The only way of mitigating these events is by improvement of the equipment or, where this turns out to be unfeasible, installing mitigation equipment. For low-power equipment a UPS is a straightforward solution, for high-power equipment and for complete installations several competing tools are emerging.

The duration of dips due to distribution system faults depends on the type of protection used. Ranging from less than a cycle for current-limiting fuses up to several seconds for overcurrent relays in underground or industrial distribution systems. The long dip duration makes that equipment can also trip due to faults on distribution feeders fed from other HV/MV substations. For deep long-duration dips, equipment improvement becomes more difficult and system improvement easier. The latter could well become the preferred solution, although a critical assessment of the various options is certainly needed.

Dips due to faults in remote distribution systems and dips due to motor starting should not lead to equipment tripping for dips down to 85%. If there are problems, the equipment needs to be improved. If equipment trips occur for long-duration dips in the 70% - 80% magnitude range, changes in the system have to be considered as an option.

For interruptions, especially the longer ones, equipment improvement is no longer feasible. System improvements or a UPS in combination with an emergency generator are possible solutions here.

The different mitigation methods are summarized in *Figure 23*: for events less severe than 70%, 100ms, the solution must be sought at the equipment or customer side. For interruptions, feasible improvements need to be made in the power system. In the intermediate area, voltage dips originating in the local distribution system, customer and distribution company together should look for a solution. Note that for the time being, both "equipment improvement" and "system improvement" may involve the installation of mitigation equipment. In the long term, the two areas need to grow towards each other. By improving the protection of distribution systems, the maximum fault-clearing time may be brought down to 100 or 200 ms. Ensuring that all equipment is immune to dips up to 100 ms, 70% would have largely solved the voltage dip problem.

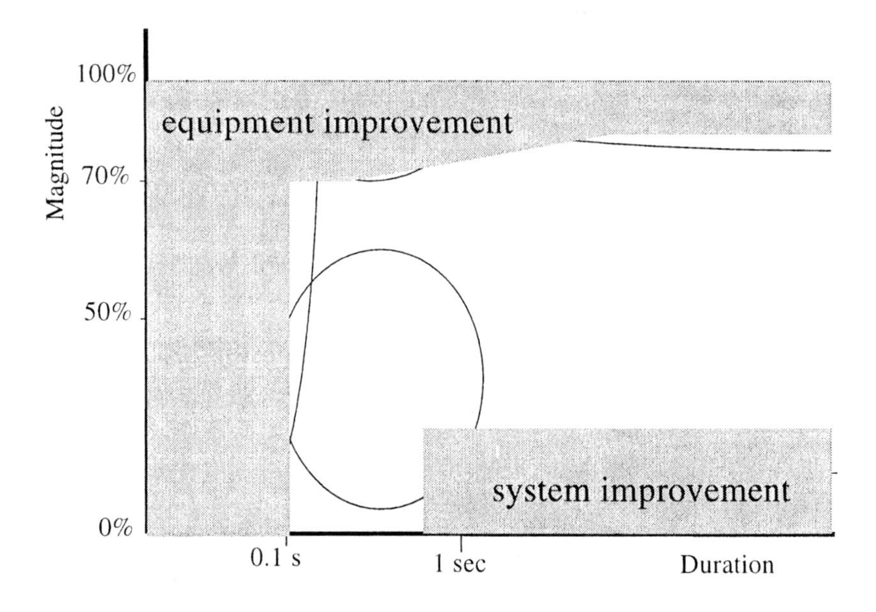

Figure 23. Different mitigation methods for different dips

3. OTHER POWER QUALITY ISSUES

Although voltage dips and long interruptions are the disturbances of most concern to customers, several other power quality issues have come up during the last few decades. These will be discussed briefly below. For each of the disturbances some thoughts are presented about among others the responsibility question. But the issues discussed below do in most cases not lead to any serious problems, so that they do not have a high priority. As long as even long interruptions and voltage dips are not yet considered in the interaction between the partners in the deregulated market, these issues may be temporarily put on hold. They should however not be forgotten and will have to be considered somewhere in the future.

3.1 Short Interruptions

Short interruptions are, according to the IEC definition, interruptions with a duration less than 3 minutes. Short interruptions are due to automatic re-closure after a fault. The practice is especially in use for overhead distribution lines, where most faults are temporary. By re-closing the line automatically, the duration of the interruption is reduced from in the order of

1 hour to often less than one minute. Seen in this light, short interruptions are an improvement of the voltage quality. However, auto-reclosure is often combined with the method of fuse-saving. An overhead distribution feeder is protected by one circuit breaker with a number of downstream fuses or disconnectors. The circuit breaker instantaneously opens for each fault; the fuses and the disconnectors isolate the fault if it turns out to be a permanent one. The result is that the customers experience many short interruptions in stead of a small number of long interruptions. The underlying philosophy is that the inconvenience due to short interruptions is small, so that the net effect will be an improvement of the voltage quality. But the same equipment that is sensitive to voltage dips is even more sensitive to short interruptions, so that the inconvenience due to short interruptions is almost as big as due to long interruptions. A survey among Swedish customers performed in 1993 [15] found the following interruption costs:

- a two-minute interruption costs 10 Swedish crowns per kW for the average customer;
- a one-hour interruption costs 34 crowns per kW for this customer;
- for the chemical industry the interruption costs are 61.6 crowns per kW for a 2-minute interruption;
- for a 1-hour interruption the costs are 87.1 crowns per kW.

Monitoring and characterization of short interruptions are much easier than for voltage dips: the only characteristic of interest is the duration. The performance indicator may consist of the interruption frequency versus the minimum duration of the event (*i.e.* the number of events longer than a given duration). Also the responsibility question is straightforward: the interruption originates with the local distribution company. Some care should however be taken with short interruptions. The performance indicators should not give the utility an incentive to simply get rid of the auto-reclosure scheme and thus increase the number of long interruptions. The incentive should be made towards reducing the number and duration of all interruptions.

The number of short interruptions can be reduced by reducing the number of faults, but also by getting rid of fuse-saving and disconnector schemes. This obviously requires the installation of more reclosing breakers. The duration of the short interruptions and possible also their number, can be reduced by introducing a distribution-automation scheme where communication between different reclosers and breakers is used to quickly determine the fault location.

3.2 Harmonic Distortion

Harmonic distortion is mainly due to the presence of non-linear load like computers, consumer electronics, adjustable-speed drives, energy-saving lighting, and electric-arc furnaces. An example of the harmonic current taken by a modern adjustable-speed drive is shown in *Figure 24*. The effect of the current distortion is a voltage distortion: the voltage waveform is no longer exactly a sine wave. The presence of capacitor banks can further amplify the harmonic distortion. The distortion can be characterized through the magnitude and phase angle of the various harmonic components (i.e. through the spectrum of the voltage or current) but for most applications it is sufficient to give the "total harmonic distortion" and the "crest factor".

Harmonic distortion mainly leads to inconvenience for distribution and transmission companies. The effects on the customer are limited. The harmonic currents lead to increased heating of series components, like cables and transformers. This in turn requires derating of those components. Derating is especially needed for transformers feeding large amounts of non-linear load. A special concern is the neutral conductor in three-phase low-voltage installations: the third harmonic currents produced by single-phase equipment add in the neutral conductor, so that the neutral current is no longer zero. In modern low-voltage installations the neutral conductor needs to be able to safely carry the same current as the phase conductors.

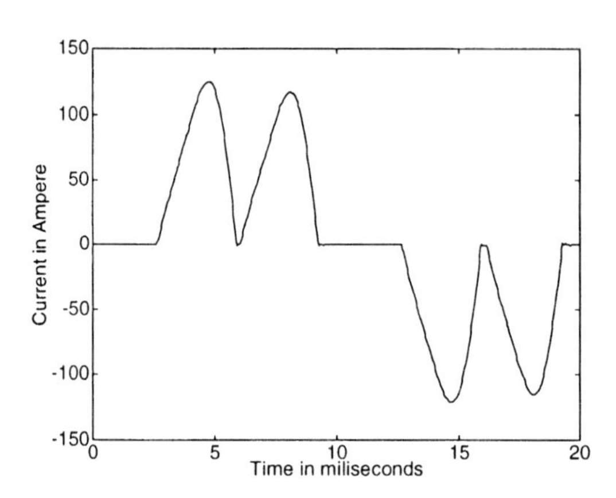

Figure 24. Example of harmonic current distortion

Customer equipment is affected by the harmonic voltage distortion. Direct equipment trips are rare and limited to cases with large or high-frequency distortion. But especially induction motors may require a certain amount of derating to allow for additional heating due to harmonic currents. Electronic equipment will experience a reduction in voltage crest factor as an actual voltage drop, leading to a reduced performance and increased sensitivity to voltage dips.

The responsibility for the harmonic distortion problem lays with both the customer and the network operators. It is the customers responsibility that current distortion does not exceed certain limits, whereas the distribution and transmission companies are responsible for the voltage distortion. Additionally some responsibility rests with equipment manufacturers, who should produce equipment causing limited current distortion. In most countries the responsibility problem is well defined in standards, like IEEE 519 in the US and IEC 61000-3-2 and 61000-3-4 elsewhere.

3.3 Transient Overvoltages

Transient overvoltages are increases in the supply voltage with a duration less than one cycle, sometimes much less. Transient overvoltages are due to switching events and lightning strokes. Switching events both in the power system and at the customer premises can lead to large overvoltages. The most well-publicized system event is the transient due to capacitor energizing. At the switching instant the voltage at the capacitor bus instantaneously drops to zero, recovering through an oscillation with the inductive part of the source impedance. This oscillation, superimposed on the fundamental frequency voltage leads to over-voltages up to 1.8pu. When capacitors are present at different voltage levels additional resonance frequencies may be excited leading to over-voltages above 2pu (a phenomenon referred to as voltage magnification).

The main direct consequence of the event is the tripping of adjustable-speed drives and sensitive electronic equipment. The relatively low frequency of the event, typically 250 to 500 Hz, makes that it propagates through the system without much damping. The long duration makes that over-voltage protection may not be able to absorb all the energy, even though the actual over-voltage is not very high. Modern synchronized switching can prevent serious over-voltages, but voltage magnification may still lead to over-voltages large enough to cause equipment mal-trips.

Surveys indicate that over-voltages of shorter duration originate mainly within the customer promises, but not enough is known about these transients to draw other general conclusions about them. This lack of knowledge also makes it all but impossible to address the responsibility question.

3.4 Voltage Variations

Voltage variations are slow variations in the rms voltage, mainly due to the daily load variations. Voltage variations are not a problem in a properly designed and operated system, but areas with fast load growth may suffer from too low rms voltages.

The responsibility for voltage variations is reasonably defined in standards. Each countries has its own limits. Sweden, for example, has been using the range 207V-247V. Within Europe the voltage characteristics standard EN50160 becomes more and more the reference for comparing voltage variations against. According to this document, the rms voltage shall not be outside of the range 207V-253V for more than 5% of the time [5]. Note that the voltage characteristics standard is strictly speaking not a requirement for the distribution companies, but a description of the existing situation. However, more and more distribution companies and customers start using the document as a requirement. When the voltage at a certain location does not fulfil the requirement, the distribution-company should improve the supply. These situations occur rarely because of the wide range allowed in most standards.

3.5 Voltage Steps

Voltage steps are sudden changes in rms voltage due to transformer tap-changer operation, load switching, and switching of shunt capacitors and inductors. Voltage steps are rarely a problem, although some processing may depend on the output of a lamp whose intensity is proportional to the square of the voltage.

The responsibility rests with the distribution or transmission company for any switching actions in their networks, and with the customers when load switching or switching actions on the customer premises lead to the voltage step. Obviously, only voltage steps exceeding a certain threshold need to be considered in any performance criterion. Again, several countries already have standards about the size of voltage steps that should be considered as normal. Voltage steps below 1 or 2 percent of the nominal voltage are probably not worth reporting.

3.6 Voltage Fluctuations

Voltage fluctuations are changes in rms voltage on a very short time scale (seconds or less) due to fast variations in the load current. Voltage fluctuations will lead to variations in the intensity of lighting, a phenomenon called *light flicker*. When the light flicker exceeds a certain severity this is interpreted by our brain as disturbing, *e.g.* leading to headaches. Well-defined standards have been developed to determine whether a certain voltage fluctuation will lead to disturbing light flicker: the so-called "flickermeter" concept. As voltage fluctuations are due to load variations, the responsibility rests with the customer behind the load variations.

3.7 Load Currents

Power quality, as introduced in Section 1.1, consists of voltage quality and current quality. In the previous sections, the discussion has centered on voltage quality issues: disturbances brought to the customer by the distribution or transmission system. The other side of the problem: disturbances brought to the network by a customer, has been touched before, *e.g.* when discussing harmonic distortion, but only when the current disturbance led to voltage disturbances.

However, current disturbances, both variations and events, may also be a problem when they do not lead to any voltage disturbance. The classical example is the reactive component of the current. This will lead to an increase in losses in the system, and requires overrating of series components like cables and transformers. Also the variations in rms current require overrating of series components. If the current was constant during the day (and during the week and during the year) the system would be able to transport much more energy.

Inrush currents due to load switching may also be a concern to the distribution or transmission system. The inrush currents are normally short enough not to lead to any thermal problems, but the setting of over-current protection is very much affected by them.

4. CONCLUSIONS

The various power quality issues are partly an extension of the reliability issues discussed in the previous chapter. This holds for phenomena such as voltage dips and short interruptions. The preferred treatment of these phenomena is the same as for interruptions: starting with the publication of performance indicators, followed by a gradual improvement of the supply. With voltage dips the improvement should partly come from the side of the equipment. Equipment should be tolerant against short and shallow voltage dips. The values of 70% and 100 ms have been suggested as reasonable limits.

Monitoring of voltage dips and short interruptions will require installation of more equipment than currently in place. Future metering equipment may include a power-quality monitoring function. Modern relays will also be able to perform this function.

Determining the location of the origin of a voltage dip is important when the responsibility question is addressed. The methodology for this needs to be further developed. Some well-proven protection algorithms can be used as a starting point for the development.

The addressing of other power quality issues will partly be treated by international standards, and will for the rest have to be included in the interaction and agreements between end-users and transmission and distribution companies.

The correct treatment of power quality issues in a deregulated environment calls for additional regulation. This is mainly due to the natural monopoly held by transmission and distribution companies.

REFERENCES

1. M. H. J. Bollen, *Understanding power quality– voltage sags and interruptions*, New York, IEEE Press, 2000.
2. M. H. J. Bollen and L. D. Zhang, "Analysis of voltage tolerance of ac adjustable-speed drives for three-phase balanced and unbalanced sags", IEEE Transactions on Industry Applications, May/June 2000, pp.904-910.
3. M. H. J. Bollen, "Fast assessment methods for voltage sags in distribution systems", IEEE Transactions on Industry Applications, Nov. 1996, pp.1414-1423.
4. D. L. Brooks, R. C. Dugan, M. Waclawiak and A. Sundaram, "Indices for assessing utility distribution system RMS variation performance", IEEE Transactions on Power Delivery, Jan. 1998, pp.254-259.
5. European standard EN-50160, Voltage characteristics of electricity supplied by public distribution systems, CENELEC, Brussels, Belgium, 1994.
6. L. Conrad, K. Little and C. Grigg, "Predicting and preventing problems associated with remote fault-clearing voltage dips", IEEE Transactions on Industry Applications, Jan. 1991, pp.167-172.
7. L. E. Conrad and M. H. J. Bollen, "Voltage sag coordination for reliable plant operation", IEEE Transactions on Industry Applications, Nov. 1997, pp.1459-1464.
8. E. W. Gunther and H. Mehta, "A survey of distribution system power quality-preliminary results", IEEE Transactions on Power Delivery, Jan. 1995, pp.322-329.
9. M. B. Hughes and J. S. Chan, "Early experiences with the Canadian national power quality survey", Transmission and Distribution International, Sept. 1993, pp.18-27.
10. IEEE Standards working group online area: http://grouper.ieee.org
11. IEEE Recommended Practice for the Design of Reliable Industrial and Commercial Power Systems (The Gold Book), IEEE Std.493-1997.
12. IEC 61000-1-1, Electromagnetic Compatibility, Application and interpretation of fundamental definitions and terms.
13. E. Styvaktakis, M. H. J. Bollen and I. Y. H. Gu, "Classification of power system events: voltage dips", Proc. of International Conference on Harmonics and Quality of Power, 2000, Florida, USA.

14. E. Styvaktakis, M. H. J. Bollen and I. Y. H. Gu, "Transformer saturation after a voltage dip", IEEE Power Engineering Letters, April 2000.
15. Avbrottskostnader för elkunder (Interruption costs for electricity customers, in Swedish), Svenska Elverksföreningen, Stockholm, Sweden, 1994.
16. L. D. Zhang and M. H. J. Bollen, "Characteristics of voltage dips (sags) in power systems", IEEE Transactions on Power Delivery, May 2000, pp.700-706.
17. L. D. Zhang, M. H. J. Bollen, "A method for characterization of three-phase unbalanced dips (sags) from recorded voltage wave shapes", Proc. of Telecommunications Energy Conference (INTELEC), 1999, Copenhagen, Denmark.

FURTHER READING

1. H. Akagi, "Trends in active power line conditioners", IEEE Transactions on Power Electronics, May 1994, pp.263-268.
2. H. Akagi, "New trends in active filters for power conditioning", IEEE Transactions on Industry Applications, Nov/Dec 1996, pp.1312-1322.
3. J. Arrillaga, D. Bradley and P. S. Bodger, *Power system harmonics*, John Wiley, London, 1985.
4. J. Arrillaga, B. C. Smith, N. R. Watson and A. R. Wood, *Power system harmonic analysis*, John Wiley, Chichester, 1997.
5. J. Arrillaga, N. R. Watson and S. Chen, *Power system quality assessment*, John Wiley, Chichester, 2000.
6. J. Arrillaga, M. H. J. Bollen and N. R. Watson, "Power quality following deregulation", Proceedings of IEEE, February 2000, pp.246-261.
7. M. H. J. Bollen, *Understanding power quality problems- voltage sags and interruptions*, IEEE Press, 1999.
8. M. H. J. Bollen, P. Wang and N. Jenkins, "Analysis and consequences of the phase jump associated with a voltage sag", Power System Computation Conference, Dresden, Germany, August 1996.
9. M. H. J. Bollen, "The influence of motor re-acceleration on voltage sags", IEEE Transactions on Industry Applications, 1995, pp.667-674.
10. M. H. J. Bollen, "Characterization of voltage sags experienced by three-phase adjustable-speed drives", IEEE Transactions on Power Delivery, October 1997, pp.1666-1671.
11. E. R. Collins and R. L. Morgan, "A three-phase sag generator for testing industrial equipment", IEEE Transactions on Power Delivery, Jan. 1996, pp.526-532.
12. T. W. Diliberti, V. E. Wagner, J. P. Staniak, S. L. Sheppard and T. L. Orfloff, "Power quality requirements of a large industrial user: a case study", IEEE Industrial and Commercial Power Systems Technical Conference, Detroit, MI, May 1990, pp.1-4.
13. R. C. Dugan, M. F. McGranaghan and H. W. Beaty, *Electric power systems quality*, New York: McGraw Hill, 1996.
14. A. E. Emanuel, "Power in non-sinusoidal situations- A review of definitions and physical meaning", IEEE Transactions on Power Delivery, July 1990, pp.1377-1383.
15. A. Greenwood, *Electrical transients in power systems*, Wiley Interscience, USA, 1971.
16. G. T. Heydt, "Electric power quality", West LaFayette, IN: Stars in a circle, 1991. Only obtainable from Stars in a circle publications, 2932 SR 26W, West LaFayette, IN 47906, USA.

17. Electromagnetic Compatibility (EMC), Part 4. Testing and measurement protocols. Section 11. Voltage dips, short interruptions and voltage variations immunity tests. IEC document 61000-4-11.
18. Recommended practice for evaluating electric power system compatibility with electronic process equipment, IEEE Std. 1346-1998.
19. W. E. Kazibwe, M. H. Sendaula, *Electric power quality control techniques*, New York: Van Nostrad Reinhold.
20. T. S. Key, "Diagnosing power-quality related computer problems", IEEE Transactions on Industry Applications, July 1979, pp.381-393.
21. M. F. McGranaghan, D. R. Mueller and M. J. Samotej, "Voltage sags in industrial power systems", IEEE Transactions on Industry Applications, March 1993, pp.397-403.
22. Measurement guide for voltage characteristics, UNIPEDE report 23002 Ren 9531.
23. van Zyl, J. H. R. Enslin and R. Spée, "Converter-based solution to power quality problems on radial distribution lines", IEEE Transactions on Industry Applications, Nov. 1996, pp.1323-1330.

APPENDIX A: IEC STANDARDS ON POWER QUALITY

IEC 61000: Electromagnetic compatibility (EMC)

Part 1: General

- Section 1: Application and interpretation of fundamental definitions and terms
- Section 2: Methodology for the achievement of functional safety of electrical and electronic equipment

Part 2: Environment

- Section 1: Description of the environment – Electromagnetic environment for low-frequency conducted disturbances and signalling in power supply systems.
- Section 2: Compatibility levels for low-frequency conducted disturbances and signalling in public supply systems.
- Section 3: Description of the environment - Radiated and non-network-frequency-related conducted disturbances.
- Section 4: Compatibility levels in industrial plants for low-frequency conducted disturbances.
- Section 5: Classification of electromagnetic environments.
- Section 6: Assessment of the emission levels in the power supply of industrial plants as regards low-frequency conducted disturbances.
- Section 7: Low-frequency magnetic fields in various environments.
- Section 8: Voltage dips, short interruptions and statistical measurement results (in preparation).
- Section 12: Compatibility levels for low-frequency conducted disturbances and signalling in public medium-voltage power supply systems (in preparation).

Part 3: Limits

- Section 1: Overview of emission standards and guides (in preparation).
- Section 2: Limits for harmonic current emissions (equipment input current ≤16A per phase).

- Section 3: Limitation of voltage fluctuations and flicker in low-voltage supply systems for equipment with rated current ≤16A.
- Section 4: Limits for harmonic current emissions. (Equipment with rated current greater than 16A per phase).
- Section 5: Limitation of voltage fluctuations and flicker in low-voltage power supply systems for equipment with rated current greater than 16 A.
- Section 6: Assessment of emission limits for distorting loads in MV and HV power systems.
- Section 7: Assessment of emission limits for fluctuating loads in MV and power systems.
- Section 8: Signalling on low-voltage electrical installations - Emission levels, frequency bands and electromagnetic disturbance levels.
- Section 9: Limits for interharmonic current emissions (equipment with input power ≤16 A per phase and prone to produce interharmonics by design) (in preparation).
- Section 10: Emission limits in the frequency range 2 ... 9 kHz (in preparation).
- Section 11: Limitation of voltage changes, voltage fluctuations and flicker in low voltage supply systems for equipment with rated current ≤75 A and subject to conditional connection (in preparation).

Part 4: Testing and measurement techniques

- Section 1: Overview of immunity tests.
- Section 2: Electrostatic discharge immunity test.
- Section 3: Radiated, radio-frequency, electromagnetic field immunity test.
- Section 4: Electrical fast transient/burst immunity test.
- Section 5: Surge immunity test.
- Section 6: Immunity to conducted disturbances, induced by radio-frequency fields.
- Section 7: General guide on harmonic distortion and interharmonics measurement and instrumentation, for power supply systems and equipment connected thereto.
- Section 8: Power frequency magnetic field immunity test.
- Section 9: Pulse magnetic field immunity test.
- Section 10: Damped oscillatory magnetic field immunity test.
- Section 11: Voltage dips, short interruptions and voltage variations immunity tests.
- Section 12: Oscillatory waves immunity test.
- Section 13: Testing and measurement techniques - Harmonics and interharmonics including mains signalling at a.c. power port, low frequency immunity tests (in preparation).
- Section 14: Voltage fluctuations – Immunity test.
- Section 15: Flickermeter - Functional and design specifications.
- Section 16: Test for immunity to conducted common mode disturbances in the frequency range 0 Hz to 150 kHz.
- Section 17: Ripple on dc input power port, immunity test.
- Section 27: Unbalance, immunity test (in preparation).
- Section 28: Variation of power frequency, immunity test.
- Section 29: Voltage dips, short interruptions and voltage variations on dc input power ports, immunity tests (in preparation).
- Section 30: Measurements of power quality parameters (in preparation).
- Section 33: Methods and means of measurements of high power transient parameters

Part 5: Installation and mitigation guidelines

- Section 1: General considerations.

- Section 2: Earthing and cabling.
- Section 6: Mitigation of external EM influences (in preparation).
- Section 7: Degrees of protection against electromagnetic disturbances provided by enclosures (in preparation).

Part 6: Generic standards
- Section 1: Immunity for residential, commercial and light-industrial environments.
- Section 2: Immunity for industrial environments.
- Section 4: Emission standard for industrial environments.
- Section 5: Immunity of apparatus for generating stations and high-voltage substations (in preparation).

Product standards

- IEC 601118-3: Hearing aids, Part 13: Electromagnetic compatibility (EMC)
- IEC 60533: Electrical and electronic installations in ships – Electromagnetic compatibility.
- IEC 61326-1: Electrical equipment for measurement, control and laboratory use - EMC requirements - Part 1: General requirements
- IEC 61543: Residual current-operated protective devices (RCDs) for household and similar use - Electromagnetic compatibility
- IEC 61547: Equipment for general lighting purposes - EMC immunity requirements
- IEC 61800-3: Adjustable speed electrical power drive systems - Part 3: EMC product standard including specific test methods
- IEC 61917: Cables, cable assemblies and connectors - Introduction to electromagnetic (EMC) screening measurements

APPENDIX B: IEEE STANDARDS ON POWER QUALITY

- Std 4-1995, Standard techniques for high-voltage testing.
- Std 120-1989, Master Test Guide for Electrical Measurements in Power Circuits.
- Std 141-1993, Recommended practices for electric power distribution for industrial plants.
- Std 142-1991, Recommended practice for grounding of industrial and commercial power systems, also known as the Green Book.
- Std 213-1993, Standard procedure for measuring conducted emissions in the range of 300 kHz to 25 MHz from television and FM broadcast receivers to power lines.
- Std 241-1990 Recommended practice for electric power systems in commercial buildings, also known as the Gray Book.
- Std 281-1994, Standard service conditions for power system communication equipment.
- Std 299-1991, Standard method of measuring the effectiveness of electromagnetic shielding enclosures.
- Std 352-1993, Guide for general principles of reliability analysis of nuclear power generating station safety systems.
- Std 367-1996, Recommended practice for determining the electric power station ground potential rise and induced voltage from a power fault.

- Std 376-1993, Standard for the measurement of impulse strength and impulse bandwidth.
- Std 430-1991, Standard procedures for the measurement of radio noise from overhead power lines and substations.
- Std 446-1987, Recommended practice for emergency and standby power systems for industrial and commercial applications, also knows as the Orange Book.
- Std 449-1990, Standard for ferroresonance voltage regulators.
- Std 473-1991 Recommended practice for an electromagnetic site survey (10 kHz to 10 GHz).
- Std 493-1997, Recommended practice for the design of reliable industrial and commercial power systems, also known as the Gold Book.
- Std 519-1992, Recommended practice and requirements for harmonic control in electric power systems.
- Std 539-1990, Standard definitions of terms relating to corona and field effects of overhead power lines.
- Std 762-1987, Standard definitions for use in reporting electric generating unit reliability, availability, and productivity.
- Std 859-1987, Standard terms for reporting and analyzing outage occurrences and outage states of electrical transmission facilities.
- Std 944-1986, Application and testing of uninterruptible power supplies for power generating stations.
- Std 998-1996 Guide for direct lightning stroke shielding of substations.
- Std 1048-1990 Guide for protective grounding of power lines.
- Std 1057-1994 Standard for digitizing waveform recorders.
- Std 1100-1992 Recommended practice for powering and grounding sensitive electronic equipment, also known as the Emerald Book.
- Std 1159-1995 Recommended practice for monitoring electric power quality.
- Std 1184-199, Guide for the selection and sizing of batteries for uninterruptible power systems.
- Std 1250-1995, Guide for service to equipment sensitive to momentary voltage disturbances.
- Std 1325-1996, Recommended practice for reporting field failure data for power circuit breakers.
- Std 1313.1-1996, Standard for insulation coordination - definitions, principles, and rules.
- Std 1346-1998, Recommended practice for evaluating electric power system compatibility with electronics process equipment.
- Std 1409 (in preparation), Custom power.
- Std 1433 (in preparation), A standard glossary of power quality terminology.
- Std 1453 (in preparation), Voltage flicker.
- Std C37.10-1995, Guide for diagnostics and failure investigation of power circuit breakers.
- Std C37.95-1994, Guide for protective relaying of utility-consumer interconnections.
- Std C37.100-1992, Standard definitions for power switchgear.
- Std C.57.110-1986, Recommended practice for establishing transformer capability when supplying nonsinusoidal load currents.
- Std C57.117-1986 Guide for reporting failure data for power transformers and shunt reactors on electric utility power systems.
- Std C62.41-1991 Recommended practice on surge voltages in low-voltage ac power circuits.

- Std C62.45-1992 Guide on surge testing for equipment connected to low-voltage ac power circuits.
- Std C62.48-1995 Guide on interactions between power system disturbances and surge-protective devices.

Index

A

Act
  Power Utilities Regulatory Policy Act of 1978 (PURPA), 53, 121
  US Energy Policy Act of 1992, 7, 122
Activities
  gencos
    bilateral markets, 96-97
    pool markets, 95
  ISO
    bilateral markets, 91-95
    pool markets, 80-91
Adequacy versus security, 230
Ageing, 207
Ancillary services, 81, 173-205
  country cases, 175-186
  what are they, 171-174
Annual peak load, role in reliability studies, 227
Australia, *see* NEMMCO, Australia
Automatic generation control (AGC) *See* frequency control
Availability, 211, 218

B

Back-up protection, 229
Balance service, Sweden, 79, 182-184
Benefits from competitive electricity market, 19
Bids,
  energy brokerage system, 62-64

probability of acceptance, 110-113
Bilateral markets, 91-97
activities of gencos, 96-97
bilateral transaction matrix, 93
role of ISO, 91-95
Black start capability service, 174, 177
Blackout, 206, 238-239, 249
British Power Pool, 11, 74-76, 83, 86-88, 233
ancillary service management, 178-179
compensation for interruptions, 243
demand-side bidding, 148
performance of distribution companies, 246
reactive power, 187-190
supply performance, 247, 248
transmission
reliability, 234
unavailability, 232
pricing, 138
Brokerage system, 61-69

C

Capacity charge, 76
congestion management, 159
Sweden, 142
Characteristic voltage, of voltage dip, 267-271
Chile, transmission pricing, 138
Circuit theory model, role in reliability analysis, 215-216
Clearing price, 76, 77, 83, 85
Common-mode failures, 230
Compensation for interruptions, 243
Competitive bidding, 107-114
hydro reserves in Norway, 109
probability of acceptance of bid, 110-113
strategies, 108
Competitive electricity market, benefits from, 19
Congestion management, 157-166
capacity fee, 159
counter trade, 165
market splitting, 159
Consumer payment minimization, 89
Contingency reserve services, 176
Cost
embedded, based transmission price, 127

incremental, based transmission price, 134
long-run marginal, based transmission price, 136
operating, transmission, 124
opportunity
reactive power supply, 192-193, 203
transmission, 124
power pool, total system, 55
reactive power production, 201
reinforcement, transmission, 124
reliability, 217, 223-226, 231, 236, 287
short-run marginal, based transmission price, 134
Counter-trade, 165
Country cases
ancillary service management, 175-186
reactive power, 187-198
generation scheduling, 82
interruptible load management, 148-151
security management, 145-146
transmission pricing, 138-142
Current quality, 254
Current-limiting coils, 283
Current-limiting fuses, 264, 266, 282, 285

D

DC power flow, 55
Demand-side bidding, 148
Demand-supply balance, 32, 49, 55, 84, 86, 101
Dependability, power system protection, 230
De-rated states, in generation modeling, 213
Deregulation
after-effects of, 21
ancillary services, 171-204
congestion management, 157-166
developing countries, 15-19
different entities in, 4-6
Europe, 9
British power pool, 10
European Union Directive, 9, 10
Nordic, 11-13
industrialized countries, 6-15, 232, 244-246
relation with reliability, 232-246
security management, 145-157
transmission open access, 121-123, 138-142

US, 7
what is, 2
Developing countries
deregulation, background, 15-19
generation reliability,228
Direct market-trading, 15
Dispatch, 29-30, 174
Distribution
companies, voltage dips, 279
reliability, 208, 231, 236
effects of deregulation, 241
regulation, 242
Double auction power pool, 83-85
Dynamic voltage restorer (DVR), 284

E

Economic load dispatch (ELD), 30-35
conditions for the optimum, 32
Lagrangian, Lagrange multipliers, 32-33
dynamic economic dispatch, 35
review of recent developments, 34
Electricity market
benefits from, 19
effect on reliability, 237-241
settlement, 75-78, 82-91
structures, 74-80
Electricity prices, relation to reliability, 225
Electromagnetic compatibility (EMC), 254
Embedded cost based transmission price, 127
EN 50160, 242, 290
Energy
economic exchange of, 54
Energy brokerage system, 61-69
Energy Policy Act of 1992, US, 7, 122
EPRI, US, 277
Equipment
role in power quality, 255, 284
voltage tolerance, 273, 275-276, 284
Europe
transmission pricing, development, 142-144
Transmission system operator (TSO), 120, 142-144
Even harmonics, 262, 264
Events, 209

power quality, 256
Expected time to failure (ETTF), 211
Exponential distribution, *see* Negative-exponential distribution

F

FACTS, 240
Failures, 205
Failure rate, 210, 211, 214
Fault-current limiter, 282
Forced
outages, 206, 212, 239
unavailability, 228
Frequency control, 172, 177, 179, 182-185
Nordel system, 184-185
primary and secondary control, 173
role of the market, 237-238
Finland, reactive power management, 197
Forecast, market price, 99

G

Generator company (genco)
pool markets, 95
bilateral markets, 96-97
voltage dips, 279
Generation
scheduling, country practices, 82
reliability, 208, 215, 227-228, 236, 249
effects of deregulation, 237-238
Growth in NUG participation, 8
Guaranteed standards of service, 244-245

H

Harmonic distortion, 288-289
Hierarchical levels, in reliability analysis, 227-231
Hydro reserves versus price in Norway, 109

I

IEC 61000-3-2, 289
IEC 61000-3-4, 289
IEC, EMC and power quality standards, 254, 294-296
IEEE 519, 289
IEEE
power quality standards, 254, 297-298

Task Force on Transmission Access, 122
Incremental cost based transmission price, 134
Independent system operator, 74
operational planning activities, 80-95
pool markets, 80-91
bilateral markets, 91-95
role in guaranteeing reliability, 237, 239-240, 242
Interconnected operation service, *see* Ancillary service
Interruption, 205-206
costs, *see* reliability costs
criterion, 207-208
Interruptible load management, 148-157
country cases
Alberta power pool, 150
Australian market, NEMMCO, 151
British power pool, 148
California, 150
New York power pool, 149
optimal incentives, 151-157
role in transmission reliability, 238-241

K

Kuhn-Tucker's conditions, 33, 84, 86

L

Lagrangian, 32, 56, 84, 86
Level-I reliability, *see* generation reliability
Level-II reliability, *see* transmission reliability
Level-III reliability, *see* distribution reliability
Life time distributions, 214
Light flicker, 290-291
Load currents, power quality phenomenon, 291
Load shedding, 237, 239
Long interruptions, *see* interruptions
Loss of load expectation, 228
Lost opportunity cost, 192-193
LRMC based transmission price, 136

M

Maintenance, *see* preventive maintenance
Marginal cost, 46
long-run, based transmission price, 136
short-run, based transmission price, 134

wheeling real and reactive power, 126
Market settlement, 75-78, 82-91
consumer payment minimization, 89
double auction, 83-85
Nordpool, 77
price forecast, 99
single auction, 85-86
UK, 76, 85-86
structures, 74-80
Market splitting, 159-165
Markov models, 217
Miniature circuit breakers, 265
Minimum cut-sets, 222-223
Monte-Carlo simulation, 207, 208, 217
Motor
-generator sets, 284
re-acceleration, cause of post-fault dip, 263-264
starting, cause of voltage dips, 261-262, 281
Multi-area joint dispatch, 55
Multi-stage voltage dips, 260-261, 275
MW-mile method, 128

N

(n-1)-criterion, 229, 239, 242
National Grid Company (NGC), *see* British power pool
Natural monopolies, 232, 240, 248
need for regulation, 241-246
Negative-exponential distribution, 214, 218
NEMMCO, Australia, 14
ancillary service management, 179-182
interruptible load management, 151
payment mechanisms, 180
reactive power, 193-195
NERC
Operating Policy 10, 175-177
role in guaranteeing reliability, 233, 240
Network models, *see* stochastic networks
Netherlands, compensation for interruptions, 243
New York power pool
reactive power management, 190-193
interruptible load management, 149
New Zealand Electricity Market (NZEM), 14
generation scheduling, 82

Non-rectangular dips, in voltage-dip co-ordination chart, 273-275
Non-utility generators, 8, 121
Nordic electricity market, 11-13, 74, 76-80
Normal distribution, 111-112, 231
NordPool, 13
  market settlement, 77
  in time-domain, 80

O

OFFER, 242, 244
OFGEM, *see* OFFER
Operating cost, transmission, 124
Operational planning, 29-30, 80-114
  activities of genco, 95-114
  activities of ISO, 80-95
  unit commitment, 47-53, 100-107
Operational reliability techniques, 236
Opportunity cost
  transmission, 124
  reactive power, New York pool, 192-193, 202
Optimal power flow (OPF), 37-47, 155-157
  basic model, 38
  security constrained economic dispatch, 45
  preventive and corrective rescheduling, 45
  reactive power planning and voltage control, 45
  power wheeling and wheeling loss calculation, 46
  interruptible load management, 151-157
Optimal reliability, 224
Outage rate, 218
Outages, 206, 209
Overall standards of service, 245-246
Overhead lines versus underground cables, 282
Overlapping outages, 206

P

Payment mechanisms, ancillary services, Australia, 180
Performance
  indicators
    interruptions, 246-248
    short interruptions, 287
    voltage dips, 271-278, 280
  standards, interruptions, 243-246
Phase-angle jumps, 276

PN-factor, of voltage dip, 267-271, 281
Point-on-wave of dip initiation, 276
Pool markets
  activities of gencos, 95
  activities of ISO, 80-91
Post-fault voltage dips, 263-264, 281
Postage-stamp method, 128
Power flow equations, 39, 155
Power pools, 53-69
  multi-area joint dispatch, 55
  total system cost, 55
  energy brokerage system, 61-69
Power quality, 208, 253-298
  contracts, 242, 272
  definitions, 253-254
  state-estimation, 278
Power Utilities Regulatory Policy Act, 7, 53, 120
Power wheeling, 121
  rate, 125, 137
Preventive and corrective rescheduling, 45
Preventive maintenance, 206, 212, 215, 228
Price forecast, 99
Pricing
  real and reactive power, 46
  power transactions, 125-137
Probability
  acceptance of bid, 110-113
  density function, 214
  distribution function, 214
Protection, contribution to reliability, 206, 210, 213, 229-231

Q

Quality of consumption, definition of, 254
Quality of supply, definition of, 254

R

Ramp rate, 34, 51
Reactive power
  ancillary service, 173, 177, 180, 186-202
  marginal cost, 46
  synchronous generator capability, 200-201
  wheeling, marginal cost, 126
Regional Transmission Organizations (RTO), 8

Regulation and load following services, 175
Reinforcement cost, transmission, 124
Reliability, 205-250
  analysis, 206-231
    effects of deregulation, 236-237
    extention towards voltage dips, 278
  cost, 217, 223-226, 231, 236, 287
    costs versus reliability curve, 223
Reliability-centered maintenance, 241
Renewal theory, 217
Repair, 210-212
Reserve services, 173
Revenue reconciliation, 137
RMS voltage, 258

S

Scheduled
  outages, 206-228
  unavailability, 228
Scheduling and dispatch service, 174
Scottish Hydro Electric, 233
Scottish Power, 233
Secondary components, 210
Security, 145-157, 208, 215
  country cases, 145-146
  interruptible load management, 148-157
  management in deregulation, 145-157
  power system protection, 230
  spinning reserves, 147
Security constrained economic dispatch, 45
Shielding wires, 282
Short circuits
  cause of voltage dips, 259-260, 264-266
  in reliability analysis, 212, 215, 230, 240
Short interruptions, 256, 286-287
Single auction power pool, 85-86
Social welfare, 82-86
Spinning reserve, 50, 147
SRMC based transmission price, 134
Static circuit breakers, 282
Static transfer switches, 284
Stochastic components, 207, 209-210
  data, 207

five-state model, 213
general model, 214
states, 209
networks, 216, 218-223
series or parallel connections, 220-221
Stuck-breaker, 210
Sweden
after-effects of deregulation, 21-25, 234, 235, 237
ancillary service management, 182-185
balance service, 79, 182-184
reactive power, 196-198
background to deregulation, 11
capacity fee, latitude dependent, 142
compensation for interruptions, 243
counter-trade, 165
transmission pricing, 139
Symmetrical components, 267
Symmetrical phase, of voltage dip, 268-271

T

Transaction
brokerage system, price, 64
point-to-point, 123
pricing, 125-137
Transmission company, transco, 120
voltage dips, 279
Transformer
energizing, cause of voltage dips, 261-262, 281
saturation, cause of post-fault dip, 263
Transient instability, 240
Transmission
capacity, 55
cost components, 123
point-to-point transactions, 123
pricing, 125-137
country cases, 138-142
embedded cost based, 127
postage-stamp method, 128
MW-mile method, 128
Europe, developments, 142-144
incremental cost based, 134
long-run marginal cost based, 136
short-run marginal cost based, 134

reliability, 208, 215, 218, 229-231, 236, 249
effects of deregulation, 238-241
Sweden, 234, 235, 238
UK, 233-234
Transmission congestion management, 157-166
Transmission open access, 121-123, 138-142

U

Unavailability, 211, 218
Uninterruptable power supply (UPS), 283, 285
Unit commitment, 47-53, 100-107
in deregulated environment, 100-107
UK, *see* British power pool
US
ancillary service management, 175-177
background to deregulation, 7
Energy Policy Act of 1992, 7, 122
Power Utilities Regulatory Policy Act, 7, 53, 120
transco, 120

V

Value of lost load, 226
Variations, power quality, 256
Voltage collapse, 238, 240
Voltage dips, 215, 257-286
co-ordination chart, 272-276
direction, 280
magnitude and duration, 258, 266
mitigation, 272, 281-286
monitoring, 277
phase-angle jump, 276
point-on-wave, 276
responsibility question, 279-281
stochastic prediction, 278
three-phase treatment, 266-271, 275-276
Voltage fluctuations, 290-291
Voltage magnitude events, 256
Voltage quality, definition of, 254
Voltage sags, *see* Voltage dips
Voltage steps, 290
Voltage variations, 290

W

Weather-related outages, 231
Weibull distribution, 217
Weibull-Markov models, 217
Wheeling, 121
  rate, 125, 137
  marginal cost, real and reactive power, 126
  revenue reconciliation, 137
  wheeling rate evaluation simulator (WRATES), 126